Progress in Physics
Vol. 3

Edited by
A. Jaffe and
D. Ruelle

Birkhäuser
Boston · Basel · Stuttgart

Yu. I. Manin
Mathematics and Physics

Translated by
Ann and Neal Koblitz

Birkhäuser
1981 · Boston · Basel · Stuttgart

Author

Yu. I. Manin
Steklov Mathematical Institute
Vavilova 42
117966, GSP-1
Moscow
USSR

2. printing 1983

Library of Congress Cataloging in Publication Data
Manin, IU. I.
  Mathematics and Physics.
  (Progress in Physics ; 3)
  Translation of: Matematika i fizika.

   1. Mathematics--1961-   . 2. Physics. I. Title.
II. Series: Progress in Physics (Boston, Mass.); 3.
QA37.2.M3413   530.1'5   81-1319
ISBN 3-7643-3027-9   AACR2

CIP — Kurztitelaufnahme der Deutschen Bibliothek
Manin, Jurij I.:
Mathematics and Physics / Yu. I. Manin.
Translated by Ann and Neal Koblitz
Boston: Basel: Stuttgart: Birkhäuser, 1981
  (Progress in Physics ; 3)
  ISBN 3-7643-3027-9
NE:GT

Originally published in Novoye v Zhizny, Naukye, Tekhnike,
Seriva Matematika, Kibernetika, Moscow: Znaniye, No.12, 1979.

©Birkhäuser Boston, 1981
ISBN 3-7643-3027-9
Pinted in USA

CONTENTS

The connection between mathematics and
physics... Is it only that physicists
talk in the language of mathematics?
It is more. This book describes how
mathematics associates to some important
physical abstractions (models) its own
mental constructions, which are far
removed from the direct impressions of
experience and physical experiment.

FOREWORD

There is a story about how a certain well-known mathematician would begin his sophomore course in logic. "Logic is the science of laws of thought," he would declaim. "Now I must tell you what science is, what law is, and what thought is. But I will not explain what 'of' means."

On undertaking the task of writing the book "Mathematics and Physics", the author realized that its size would hardly be sufficient to attempt to explain what the "and" means in the title. These two sciences, which were once a single branch on the tree of knowledge, have by our time become rather separate. One of the reasons for this is that in this century they have both engaged in introspection and internal development, i.e., they have constructed their own models with their own tools. Physicists were disturbed by the interrelation between thought and reality, while mathematicians were disturbed by the interrelation between thought and formulas. Both of these relations turned out to be more complicated than had previously been thought, and the models, self-portraits, and self-images of the two disciplines have turned out to be very dissimilar. As a result, from their earliest student days mathematicians and physicists are taught to think differently. It would be wonderful to master

both types of professional thinking just as we master the
use of both a left and a right hand. But this book is like
a melody without accompaniment.

The author, by training a mathematician, once delivered
four lectures to students under the title "How a mathematician should study physics". In the lectures he said that
modern theoretical physics is a luxuriant, totally Rabelai-
sian, vigorous world of ideas, and a mathematician can find
in it everything to satiate himself except the order to which
he is accustomed. Therefore a good method for attuning oneself to the active study of physics is to pretend that you
are attempting to induce this very order in it.

In the book, which evolved from these lectures and later
reflections, I have tried to select several important
abstractions of the two sciences and make them confront one
another. At the very highest level such abstractions lose
terminological precision and are capable of becoming cultural
symbols of the time: we recall the fate of the words "evolution", "relativity", or "the subconscious". Here we go
down to a lower level and discuss words which, although not
yet symbols, have passed beyond being merely terms: "set",
"symmetry", "space-time". (Cf. M.M. Bakhtin's attempt to
introduce this last concept into literary criticism through
the deliberately foreign term "chronotope".) Some of these
words occur in the titles of chapters. Each reader must
have in his or her mind preliminary images of these concepts,
images which have a physical origin in the broad sense of the
word.

The author wishes to show how mathematics associates new mental images with such physical abstractions; these images are almost tangible to the trained mind but are far removed from those that are given directly by life and physical experience. For example, a mathematician represents the motion of the planets of the Solar System by a flow line of an incompressible fluid in a 54-dimensional phase space, whose volume is given by the Liouville measure.

The reader may need an effort of will to perceive mathematics as a tutor of our spatial imagination. More frequently one associates mathematics with rigorous logic and computational formalism. But this is only discipline, a ruler with which we are being taught not to die.[†]

The computational formalism of mathematics is a thought process that is externalized to such a degree that for a time it becomes alien and is turned into a technological process. A mathematical concept is formed when this thought process, temporarily removed from its human vessel, is transplanted back into a human mold. To think... means to calculate with critical awareness.

The "mad idea" which will lie at the basis of a future fundamental physical theory will come from a realization that physical meaning has some mathematical form not previously associated with reality. From this point of view

---

[†]An allusion to Boris Pasternak's poem "To Briusov", in which the poet says: "There was a time when you yourself each morning/ Taught us with the strict beat of your ruler how not to die." (Translator's note)

the problem of the "mad idea" is a problem of choosing, not of generating, the right idea. One should not understand this too literally. In the 1960's it was said (in a certain connection) that the most important discovery of recent years in physics was the complex numbers. The author has something like that in mind.

I do not wish to apologize for the subjectivity of the opinions and choice of material. Physics and mathematics have been written about by Galileo, Maxwell, Einstein, Poincaré, Feynman, and Wigner, among others; only the hope of saying something subjective of one's own can justify a new attempt.

1. A BIRD'S EYE VIEW OF MATHEMATICS

Mathematical Truth

The simplest mathematical operations are arithmetical computations of the following sort:

$$\frac{0.25}{20} \times \frac{\sqrt{13}}{1.1} \times \frac{7.8 \times 10^4}{2.04 \times 10^5} \times \frac{2 \times 0.048}{0.021 + 0.019} = 0.038 \ .$$

For verisimilitude this example was not copied from a student's workbook, but from a paper by Enrico Fermi and E. Amaldi "On the absorption and the diffusion of slow neutrons". Let's think a little bit about what this computation means.

a) To check this equality we can agree that it relates to integers (by squaring, clearing the denominators, and taking our unit to be a thousandth part of one). Our equality can then be considered as a prediction about the result of some "physical experiment", namely the following: take two groups of 48 objects (2×0.048), repeat this 78000 times ($\times \ 7.8 \times 10^4$), etc. In such a way first-graders make little blocks into piles in order to learn the meaning of "counting", "integer", "addition" and "multiplication", and also the meaning of arithmetic identities. Thus, it is reasonable to think of integer arithmetic as the "physics of collecting objects into piles".

1

b) Nevertheless, practical computation is, of course, carried out differently; it consists of a series of certain standard transformations of the left hand side of the identity. We take a group of symbols on the left, say $0.25 \div 20$, and replace it by 0.0125 using schoolbook recipes, and so on. All the rules, including the rule about the order of operations, can be stated beforehand. The infallibility of a computation lies in its grammatical (prescriptive) correctness; it is this which ensures the "physical truth" of the result. (Of course, Fermi and Amaldi round off the left hand side; it is clear even without calculations that their equality cannot be literally true, since $\sqrt{13}$ is an irrational number.)

c) For Fermi and Amaldi the meaning of this computation can be summarized in the following way: "We found that the narrowest band is group A (radiation strongly absorbed by silver), for which $\Delta w/w = 0.04$ ... the probability that a neutron remains in the same group after one free path is fairly low." (The 0.04 is the 0.038 on the right hand side.) Clearly we cannot immediately arrive at this conclusion, no matter how we may imagine the meaning of an arithmetical computation. Neither the laying out of 78000 piles with 96 objects in each, nor the division of 0.096 by 0.04 has in itself any relation to neutrons. Mathematical reasoning enters into the physics article along with the act of giving it a physical interpretation; it is this act which is the most striking feature of modern physics.

Be that as it may, even our simple example indicates three aspects of mathematical truth, which can conditionally be designated as *set-theoretic truth*, *formal deducibility* or *provability*, and *adequacy of the physical model*.

For mathematics, as long as it is self-contained, only the first two aspects are essential, and the difference between these two aspects was not understood until the twentieth century. Consider an assertion that can be stated simply, such as the conjecture of Fermat. Although we do not know how to prove it, or to refute it, we can be sure that it is either true or false. This confidence is based on the abstract possibility of performing infinitely many arithmetic operations (or "laying out piles"), in this case compiling all sums of powers of pairs of integers. In general, the notion of the truth of (most) mathematical assertions relies upon being able to imagine such infinite series of tests. However, every mathematical proof, i.e., argument consisting of successive applications of axioms or logical rules of deduction, is an essentially finite procedure. In the 1930's Kurt Gödel proved that for this reason provability is significantly narrower than truth, even when one only talks about the integers. In this connection it does not matter what axioms we start from, as long as they are true and are given by a finite list (or by a finite number of rules for generating them). This difference between truth and provability is widely known nowadays, but it seems that its consequences are poorly understood. Much is written discussing the problem of reductionism: Does biology or chemistry reduce to physics? It is clear that in this context one can speak only about some theoretical model of physical, biological or chemical phenomena, which is, moreover, sufficiently mathematicized. But then one must explain what is meant by reducibility-- whether it is an abstraction of a type of truth or of a type of deducibility from axioms. Thinking through both possibil- ities, we are struck with the impression that, when speaking about the reducibility of laws, we simply do not understand what we are talking about.

Sets

Modern notions of mathematical truth are connected with the development of two major conceptions: set-theoretic mathematics and the mathematics of formal languages. Every- one is familiar with some mathematical formalism. The mathe- matical baggage of the student typically may consist of the ability to carry out arithmetic operations with numbers in the decimal system, transform algebraic identities, differ- entiate and compute some integrals. This language of calculus, which had already been essentially developed by the time of Euler and Lagrange, has turned out to be very convenient, effective for the solution of problems, and accessible for mass study. Parallel to this language there has been a developing understanding of what this language talks about, i.e., the meaning of such concepts as $\sqrt{-1}$, function, differ- ential, etc. has been elucidated. Here an immense role was played by geometric representations: complex numbers lost their mystery only after Argand and Gauss proposed a coherent interpretation of them by points in the Euclidean plane; differentials were eventually interpreted using the notion of

a tangent space, and so on. Set-theoretic concepts have
lain a universal foundation for defining all mathematical
constructions using some such "generalized geometric" image.
These mental constructs represent both a receptacle for the
meaning of mathematical formalisms and a means for extracting
the meaningful assertions from the vast sea of derivable
mathematical formulas.

In this book we wish to demonstrate the value of such
mental constructions as an intermediary between mathematics
and physics. Of course, the possibilities for giving a
popular presentation of these constructions are limited. In
99 pages we cannot explain their exact meaning or teach the
reader how to use them to solve problems. But perhaps some
ideas of mathematics and theoretical physics will become
clearer. ˙The difficulty of understanding the concepts of
quantum theory or general relativity theory is in part connected with the fact that attempts to explain them omit this
act of giving an intermediate set-theoretic interpretation of
the mathematical models. Even in university teaching, in-
sufficient attention is payed to this; communication between
a physicist and a mathematician is frequently complicated by
the physicist's inclination to pass from formulas directly to
their physical meaning, omitting the "mathematical meaning".
It should be noted, incidentally, that the situation has
noticeably improved in recent years.

A good physicist uses formalism as a poet uses language.
He justifies the neglect of the commands of rigor by an
eventual appeal to physical truth, as a mathematician cannot

permit himself to do. The choice of a Lagrangian in the unified theory of weak and electromagnetic interactions of Salam and Weinberg, the introduction of Higgs fields in it, the subtraction of vacuum expectation values and other sorcery, which leads, say, to the prediction of neutral currents--all this leaves the mathematician dumbfounded.

But we return to sets.

The most important sets for physicists are *not* sets of *objects*, but *sets of possibilities*: the configuration space of a system is the set of its possible instantaneous states, space-time is the set of possible events going off like "flashbulbs", each of which gives a space-time point. A physicist is usually in a hurry to introduce coordinates in this space, i.e., functions with numerical values. If a set of n such functions allows one to coordinatize the points of the set in a unique way, then one may suppose that this set lies in n-dimensional real space R^n, which consists of vectors of the form $(a_1,\ldots,a_n)$. "Figures", i.e., subsets of this space; measurements of distances, angles, volumes, etc. in it; and, finally, its motions or its mappings into itself, all make up the physicist's main arsenal of geometric forms. It is important here that the dimension n can be as large as desired, even infinite. In a rigorous mathematical text the definition of infinite dimensionality must be introduced separately, but here we shall imagine infinite dimensionality as being "indefinitely large finite dimensionality". If the coordinates take complex values, then our sets are in C^n. But in general it is not necessary to dwell upon coordinates.

Physically they are often the vestiges of oversimplified representations of an observation; mathematically they are a
reminder of a time when there was no language in which one
could meaningfully discuss sets other than sets of numbers or
vectors.

Set-theoretic language is good in that it does not force
one to say anything superfluous. Georg Cantor defined a set
as "any collection into a whole of definite and separate
objects of our intuition or our thought". This is the best
way to explain a set as a concept which aids in the perception of the world.

Higher-Dimensional Space and the Idea of Linearity

If a car moves twenty meters in a second, then in two
seconds it will probably have gone forty meters. If a weak
wind deflects a light bullet by three centimeters, then a
wind that is twice as strong will deflect it by six centimeters. The response to small perturbations depends linearly
on the perturbation--this is a principle of natural science
which lies at the basis of a vast number of mathematical
models. A mathematician turns this principle into the definition of a differentiable function and into the postulate
that most processes are described by such functions most of
the time. Do Hooke's law of elasticity and Ohm's law say
something more than this principle of linear response to
small forces? The answer is yes if it turns out that the
laws still remain valid for somewhat larger perturbations.

Linear space is an idealization of "arbitrarily large small perturbations". One need not introduce coordinates; it is only necessary to keep in mind that the elements of a linear space (vector space) can be added together and multiplied by numbers (real or complex--this epithet is added to the name of the space). The archetypical geometric construct is our "physical space" of dimension three; the spaces R^n and C^n with coordinate-wise addition and multiplication by numbers exhaust all the finite dimensional vector spaces.

The dimension of a vector space is the number of independent linear coordinate functions on it. There is a theorem that says that the dimension does not depend on the choice of these coordinate functions. For all its simplicity, this theorem is a deep result. It establishes the first connection between something continuous and something discrete: an integer--the dimension--appears for the first time not as the number of objects or discrete forms, but as a measure of the size of a continuous entity.

A linear mapping or operator is the idealization of a linear response to arbitrary perturbations. The response can be measured by elements of the same space as the perturbation or by elements of another space; in any case, it is a mapping from a vector space to a vector space which takes a sum of vectors to the sum of their images and the product of a vector by a number to the product of the image by the same number.

Every linear mapping of a one-dimensional space into itself is multiplication by some number--the "magnification coefficient". In the complex case the geometric form is a bit more complicated: since one-dimensional complex space is viewed as a real plane, multiplication by a complex number is a combination of a real dilatation and a rotation. Pure rotations, i.e., multiplication by numbers of modulus one, play a large role in quantum mechanics: the law of evolution of an isolated quantum system is formulated in terms of such rotations.

An important class of linear mappings of an n-dimensional space into itself consists of dilatations in n independent directions with a corresponding magnification coefficient in each direction. The set of "coefficients of dilatation" of a linear operator is called its spectrum: the homonymy with a term from physics reflects a deep connection.

In quantum physics the idea of linearity acquires a fundamental physical meaning thanks to the basic postulate on superposition of quantum states. In classical physics and mathematics, in addition to the original idea that "anything" can be linearized on a small scale, a large role is played by the observation that functions on any set (all of them, or only continuous ones, or differentiable functions, or Riemann integrable functions, etc.) form a vector space, because they can be added to each other and multiplied by numbers. Most function spaces are not finite dimensional, but the possibility of systematically learning to apply to them our originally developed finite dimensional (even three-dimensional) intuition has turned out to be an extremely fruitful discovery.

This was taught to us in the twentieth century by David
Hilbert and Stefan Banach.

Measurements in a Linear Space

In three-dimensional physical space there exist rigid
bodies that preserve some "self-identity" in large space-time
domains. This is the basis of all physical measurement. It
is not at all obvious beforehand which of the idealized
properties of physical measurement will be most useful in
mathematical theory and in applications. In fact, the mathematical notions that are connected with ideas of classical
measurement form a complex conceptual design. We can immediately name some of the concepts: length, angle, area,
scalar product, motion.

So that the reader does not get a false impression, we
remark that the concept of a vector space by itself contains
nothing that allows us to measure anything in a unique way.
Vectors have no length (although proportional vectors have a
natural ratio of lengths), there is no natural measure of the
angle between two vectors, and so on. Therefore, for a mathematical formulation of the idea of measurement we must introduce an additional geometric concept, or even several concepts; to use mathematical jargon, we have to provide the
space with additional structure.

The first of these concepts is the unit sphere of the
space: the set of vectors of unit length. If any nonzero
vector lands on the unit sphere after we multiply it by some

number a, we can write down the length of the vector, which
will be $|a|^{-1}$; hence, a should be uniquely determined up to
multiplication by a number of modulus one. The distance
between two vectors x and y can be defined as the length
of their difference $|x-y|$. If nonzero vectors have nonzero
length, if the so-called triangle inequality holds: $|x+y| \leqslant$
$|x| + |y|$, and if in addition we have a condition guaranteeing
the existence of limits of Cauchy sequences, then we are led
to the concept of a Banach space. Many useful function
spaces are Banach spaces.

An arbitrary Banach space, however, is not sufficiently
symmetric to be considered the correct generalization of the
unit sphere in three-dimensional space. There are two
different methods for obtaining the necessary symmetry:
(a) require that the sphere be taken into itself by some
$n(n-1)/2$-dimensional continuous group of linear mappings
(where n is the dimension of the space); (b) require that
any two vectors in the space have a scalar product (x,y)
(this is a linear function with respect to each argument in
the real case and is a little more complicated in the complex
case) such that $|x|^2 = (x,x)$ for all x. The first method
is a generalization of the idea that rigid bodies can rotate,
while the second method is a generalization of the idea that
one can measure the angle between two vectors in a way that
does not vary under rotations of the pair of vectors. Both
ideas are closely connected to one another, and lead to the
concept of higher-dimensional Euclidean space (in the complex
case it is called Hilbert space). In suitable coordinates

the unit sphere in such a space is given by the usual

equation $\sum_{i=1}^{n} |x_i|^2 = 1$. Rotations are linear mappings from

this sphere into itself; they form a group which is denoted

$O(n)$ in the real case and $U(n)$ in the complex case. In

real Euclidean space the scalar product takes real values and

is symmetric: $(x,y) = (y,x)$. In complex Euclidean space it

takes complex values, and we have to take the complex con-

jugate when the vectors are transposed: $(x,y) = \overline{(y,x)}$. In

both cases we have the important inequality $|(x,y)|^2 \leqslant$

$|x|^2 \cdot |y|^2$, so that the number $(x,y)/|x| \cdot |y|$ is no greater

than one in modulus. In a real space this number is real,

and so there exists an angle ϕ for which $\cos\phi = (x,y)/|x| \cdot$

$\cdot |y|$. In the complex case this angle can be defined by:

$\cos\phi = |(x,y)|/|x| \cdot |y|$. The right hand side here only takes

values lying between zero and one, and we note that there is

a remarkable physical quantity with the same property--a

probability. In quantum mechanics the numbers $\cos^2\phi$ are

interpreted as probabilities--we shall discuss this in more

detail below. In high school geometry two vectors x and y

are said to be orthogonal if the cosine of the angle between

them is equal to zero, i.e., $(x,y) = 0$; the same terminology

is used in the general case as well.

If we abandon some of the properties of being Euclidean,

then the concept of scalar product can lead to several new

important classes of linear geometries. For example, in R^4

one can give the "length" of a vector $x = (x_0, x_1, x_2, x_3)$ by

the formula $|x|^2 = x_1^2 + x_2^2 + x_3^2 - x_0^2$. The one minus sign leads

to many differences from the Euclidean case: for example,

there are whole lines consisting of vectors of length zero. They represent light rays in the basic space-time model for the special theory of relativity--the famous Minkowski space.

If we reject the condition $(x,y) = (y,x)$ and replace it by $(x,y) = -(y,x)$, then every vector in this type of space will be "orthogonal to itself"! This geometry, which is called symplectic geometry, must be studied for a long time in order to become accustomed to it. The gyroscope that guides a rocket is an emissary from a six-dimensional symplectic world into our three-dimensional one; in its home world its behavior looks simple and natural. Although symplectic geometry was discovered in the last century, its role in physics was underestimated for a long time, and in textbooks this role is still obscured by the old formalism.

But it is time now to return to the Euclidean world, although in higher dimensions. The last thing we wish to discuss is the measurement of volume. If $e_1, \dots, e_n$ are mutually orthogonal vectors of unit length, then the n-dimensional unit cube spanned by them is the set of vectors of the form $x_1 e_1 + \dots + x_n e_n$, where $0 \leqslant x_i \leqslant 1$. It is natural to stipulate that its volume is equal to one. Translating this cube by any vector does not vary its volume; the cube with edge of length a is then given volume a^n. After this the volume of any n-dimensional figure can be defined by paving it over by a large number of small cubes and adding their volumes. Problems crop up around the boundary, where a bit of free space still remains, and in attempting to pave this over the cubes begin to go outside the boundary. But if the boundary is not very jagged, then, by dividing the cube into still smaller ones, we can make the error as small as we wish. This is the basic idea of integration.

We complete our construction of integration as follows. Suppose that in our area of space there is something, a "substance" as was said in the last century, that is characterized by its density f(x) close to the point x. The total amount of this substance will be approximately equal to the sum of the amounts of it in all the cubes of the paving, and the amount in one cube will be the product of the volume of the cube and the value of the density at some point in its interior. The whole sum is the "Riemann sum", and its limit is the integral of the function f over the volume.

In mathematics it is difficult to find a more classical and yet a more pertinent concept than the integral. Every few decades brings new mathematical variants of it, and physics requires more all the time. The definition of the Riemann integral, which we gave above, is mathematically reasonable only for functions f that do not vary too rapidly, such as continuous ones. But almost every physical model, as soon as it is replaced by a more detailed one, reveals that the function f, which appeared rather smooth, is in fact the result of averaging a more complicated "fine-grained" picture. Charge can be measured by integrating its density, until we reach a scale where the carriers of the charge are electrons and ions. The density of the charge on a point carrier is infinite, while off the carrier it is zero, and we are forced to construct an apparatus for integrating such functions.

The remarkable path integrals of Feynman still remain a challenge to mathematicians. These integrals have already become a fundamental instrument of quantum field theory, but they are not yet well defined as mathematical objects. Two circumstances get in the way of understanding them: the integration must be carried out over an infinite-dimensional space, and, moreover, it is strongly oscillating functions that must be integrated. Here we will say a few words about the effects of infinite dimensionality, which we can understand naively as very large finite dimensionality.

From a segment of length 1 we take out a middle part of length 0.9. The length of the remainder is, of course, 10% of the length of the whole segment. From a disc of diameter 1 we extract a concentric disc of diameter 0.9. The area of the annulus that remains is now 19% of the area of the disc. From a ball of diameter 1 we remove a concentric ball of diameter 0.9. The volume of the remaining shell is 27.1% of the volume of the ball: almost a third, as compared with one-tenth for the segment. It is not hard to see "physically" that the volume of an n-dimensional ball of diameter d should be expressed by the formula $c(n)d^n$, where $c(n)$ is a constant which does not depend on d. The fraction of the volume occupied by a concentric ball of diameter 0.9d will therefore be $(0.9)^n$; it tends to zero as n increases. Almost two-thirds of a twenty-dimensional watermelon with radius 20 cm consists of rind if the rind is 1 cm thick:

$$1 - \left(1 - \frac{1}{20}\right)^{20} \approx 1 - e^{-1} , \qquad e \approx 2.72 .$$

These calculations lead to a geometric principle: "The volume of a multidimensional solid is almost entirely concentrated near its surface." (It is interesting to take a cube instead of a ball--the same effect appears as the number of faces increases.)

Let us imagine the simplest model of a gas: N point atoms move in a reservoir with velocities v_i, $i = 1, \ldots, N$; each atom has mass m. The kinetic energy E of the gas is equal to $\sum_{i=1}^{N} mv_i^2/2$; the state of the gas described by a set of velocities with fixed total energy E determines a point on the (N-1)-dimensional Euclidean sphere of radius $\sqrt{2E/m}$. For a macroscopic volume of gas under normal conditions, the dimension of this sphere is of order $10^{23} - 10^{25}$ (determined by Avogadro's number), i.e., it is very large. If two such reservoirs are joined so that they can exchange energy but not atoms, and the sum of their energies $E = E_1 + E_2$ remains constant, then most of the time the energies E_1 and E_2 will be close to the values that maximize the volume of the space of states attained by the combined system. This volume is equal to the product of the volumes of spheres of radii $\sqrt{2E_1/m}$ and $\sqrt{2E_2/m}$, the first of which grows very rapidly as E_1 increases, while the second rapidly diminishes. Therefore, their product has a sharp peak at a point which is easily computed; the point corresponds to the condition that the temperatures be equal. The "concentration of the volume of a multidimensional solid near its surface" in essence ensures the existence of temperature as a macroscopic variable.

What space are we talking about in this example? The space of states of a physical system, more precisely, some quotient space of it (since we do not take into account either the positions of the atoms or the directions of the velocity vectors). A point of this space is again a possibility. A typical set is not the set of chairs in a room, nor the set of pupils in a class, nor even the set of atoms in a reservoir, but rather the set of possible states of the atoms in the reservoir.

The asymptotic properties of multidimensional volumes form the geometric arsenal of statistical physics. Nature constructs all of the variety in the world from a small number of different bricks. Bricks of each type are identical, and when statistics describes the behavior of conglomerates of them, it uses the concept of a point wandering in domains of an almost infinite dimensional phase space. Macroscopic observations allow us only roughly to indicate a region which contains the point at any moment in time; the larger the volume of the region, the more probable it is that we actually see the point there.

In infinite dimensions, where almost all the domain is its boundary, in order to find valid rules of thinking and computing we need some of the most refined tools of the mathematical arsenal.

Nonlinearity and Curvature

Just as the idea of linearity extrapolates small incre-
ments, the idea of curvature uses such an extrapolation to
study the deviation of a geometric object (for example, the
graph of a function f) from a linear object.

Small dimensions nourish the intuition that engenders
the geometric conception of curvature. The graph of the
curve $y = ax^2$ in the real plane has three basic forms: a
"cup" (convex downward) for $a > 0$, a "dome" (convex upward)
for $a < 0$, and a horizontal line for $a = 0$. The number a
determines the steepness of the sides of the cup or dome, and
also the radius of curvature at the lowest (highest) point;
this radius of curvature is equal to $1/(2|a|)$. In physical
models of small oscillations a heavy ball rolling in the
bottom of a cup goes through the same oscillations as an
object on a spring, and the equivalent curvature is expressed
through the mass of the ball and the rigidity of the spring.
In higher dimensions the graph of a quadratic function takes
the form $y = \sum_{i=1}^{N} a_i x_i^2$ in a suitable coordinate system.
Among the coefficients there can be positive numbers, negative numbers and zeros; they determine the number of directions in which the graph goes upward, goes downward, or
remains horizontal. In modern quantum field theory, this
simple model gives the appearance of the spectrum of masses
of elementary particles and the spectrum of forces (constants)
of interaction in an astonishingly large number of situations;
it serves as a first step on the long path to more refined
schemes. There quadratic functions arise in an infinite

dimensional space, and horizontal directions in the graphs appear because of the action of symmetry groups. When a function does not change under certain motions of the space into itself, its graph cannot be a pit, but it can possibly have the form of a ravine (a ravine can be translated into itself along the bottom, but a pit cannot).

Thus, the first notion of nonlinearity, which we have briefly described, is to ask how far a multidimensional surface--the graph of a function--deviates from a linear surface near a point at which this linear surface is tangent. Imagining the tangent space to be horizontal, we can distinguish a set of mutually orthogonal directions in it, along each of which the surface goes upward, goes downward, or remains horizontal; the speed with which the surface goes upward or downward is measured by the radius of curvature; there are as many of these radii of curvature as there are dimensions of the surface.

To describe the deviation we thus use an "external mold". This circle of ideas is natural and useful, but Einstein's theory of gravitation and Maxwell's theory of electromagnetism, and also, as we are now beginning to understand, the theory of nuclear forces and, perhaps, all interactions in general require finer notions of curvature. The first mathematical theories of "intrinsic curvature" (as opposed to the previously described "extrinsic curvature") were developed by Gauss and Riemann.

The concept of intrinsic curvature is constructed first for a domain in numerical space, in which for each pair of nearby points one is given a distance between them. The geometry of our domain with the new Riemannian concept of distance should be Euclidean on an infinitesimal scale. But various aspects of the concept of curvature show the extent to which this geometry is different from the flat Euclidean geometry on a large scale.

In order to explain these aspects it is convenient to start with the analog of lines in this space ("manifold")-- these are the geodesics, the lines of least length joining points of the space (for example, arcs of great circles on a sphere). The length of a curve is, of course, measured by an integral: the curve must be divided into many small segments, and the length of each segment is approximately measured by the distance between its endpoints.

Now we imagine a small vector in the manifold, moving along a geodesic curve so that its angle with the geodesic always remains constant. (On a two-dimensional surface this recipe determines a motion uniquely, but in higher dimensions we need more information to specify the motion precisely.) Since the space is almost Euclidean on a small scale, it is not difficult to give this prescription a precise meaning. Such a motion of a vector is called a parallel translation. We can define parallel translation along any curve: as in the computation of length, the curve should be replaced by a polygonal line composed of short segments of geodesics, and then the vector should be translated in a parallel fashion along these geodesics.

We consider a small closed curve in space--an almost
flat loop. After translating a vector along this loop and
returning to the starting point, we discover that the vector
has rotated by a small angle from its original position; this
angle is proportional to the area of the loop. Moreover, the
coefficient of proportionality depends upon: (a) the point
about which the loop is situated; (b) the direction of the
two-dimensional area spanned by the loop. This coefficient,
as a function of the point and the two-dimensional direction
at the point, is called the Riemann curvature tensor. For
flat Euclidean space the curvature tensor is identically zero.

It took a long time to see which parts of this construction are the most important, and to come to the conclusion
that the idea of parallel translation along a curve is the
fundamental one. The geometric picture of curvature that is
most relevant for understanding, for example, the theory of
Yang-Mills fields in modern physics is more general than the
picture of Riemannian curvature. To define Riemannian curvature we translated a vector along a curve in space. Now
imagine this vector as a small gyroscope, and the curve as
its world-line in four-dimensional space-time (this will be
discussed in more detail below, in the chapter on space-time).
Suppose that two gyroscopes in the same initial state diverge
and are then brought together for comparison. Then a prediction of how the directions of their axes differ is a
straightforward task of physics. Meanwhile, the set of
behaviors that can be imagined for all possible gyroscopes
of this type is a mathematical concept; it is a space

supplemented by rules for parallel translation of tangent
vectors along curves.

Yet one step remains before we can introduce the general
mathematical concept of a space with a connection and the
curvature of the connection. The direction of the axis of
the gyroscope is a special case of the notion that a localized
physical system can still possess internal degrees of freedom.
In classical physics this is an idealization, as a consequence
of which we summarily take into account the composite parts of
the system, their rotations, oscillations, etc.; in quantum
physics one has nonclassical degrees of freedom, such as the
spin or magnetic moment of an electron, which do not reduce
to imaginable behavior of the "parts" of an electron in space-
time. In general, suppose we are given a pair of spaces M
and E and a mapping $f:E \rightarrow M$, where, say, M is a model of
space-time and at each point m of M there is a localized
physical system with space of internal states $f^{-1}(m)$. Then
a connection on this geometric object is a rule for translating the system along curves in M. In other words, if we
know a piece of the world-line of a system in M and the
initial internal state of the system, then, using parallel
translation according to the connection, we know its whole
history. The curvature of a connection measures the difference between final states of the system, if we go from nearby
initial points of space-time to nearby final points by
different paths, with the systems in the same state at the
beginning.

These ideas connect the geometry directly with the physics; we are omitting the complicated turns of brilliant guesswork, error, formalism and historical accident which accompanied the emergence of these new ideas and which are constantly retold in textbooks.

A gravitational field is a connection in the space of internal degrees of freedom of a gyroscope; it controls the evolution of the gyroscope in space-time. An electromagnetic field is a connection in the space of internal degrees of freedom of a quantum electron; the connection controls its evolution in space-time. A Yang-Mills field is a connection in the space of colored internal degrees of freedom of a quark.

This geometric picture now seems to be the most universal mathematical scheme for the classical description of an idealized world in which a small number of basic interactions is considered in turn. Matter in space-time is described by a cross-section of a suitable fibre bundle $E \to M$, indicating the state in which this matter is found at each point at each moment. A field is described by a connection on this bundle. Matter affects the connection, imposing restrictions on its curvature, and the connection affects the matter, forcing its "parallel translation" along world lines. The famous equations of Einstein, Maxwell-Dirac and Yang-Mills are exact expressions of these ideas.

But even without writing down the equations, we can say a lot. The discovery that the fundamental physical fields are connections was neither as dramatic nor as easy to date

precisely as the discovery of these equations. In the theory
of gravitation, for example, the basic concept for Einstein
was not a connection, but a (pseudo-) Riemannian metric in
space-time. It was Hermann Weyl who first proposed that the
electromagnetic field is a connection, although in pre-quantum
physics he could not correctly indicate the fibre bundle on
which this connection defines parallel translation. Rather,
he guessed that the field varies the lengths of segments
passed along different paths in space-time. Einstein indicated the unsuitability of this view, and the correct bundle
on which the Maxwell connection acts was discovered by Dirac.
But it took so long to realize fully the physical peculiar-
ities of the connection field that it was only in the 1960's
that Aharanov and Bohm proposed an experiment which truly
showed the "connection-like" behavior of the Maxwell field.
For this, one has to divide an electron beam into two parts,
allow these two parts to go around a cylindrical region containing a magnetic flux, and then observe the interference
pattern on a screen. According to their assumption, the
picture on the screen should change after the magnetic field
is turned on and off, even though the beams always pass
through regions where the electromagnetic field strength is
zero when going around the domain of the magnetic flux. Thus,
the beams that are divided and rejoined on the screen "feel"
the curvature of the connection on the Dirac bundle which
comes from turning on the field in the domain they go around.
The interference pattern on the screen in fact reflects the
difference of the phase angles in the spin space of the

degrees of freedom of the electron; this difference appears
because electrons arrive at a point of the screen by different
paths in going around the magnetic flux. (The experiment was
actually carried out and corroborated these predictions.)

Some Novelties

The "gallery" of important geometric shapes through
which we hastily shepherded the reader is by no means ex-
hausted by these exhibits. The number of such concepts is
constantly increasing. Among those that have attracted the
attention of physicists and mathematicians in recent years
one can name, for example, "catastrophes", "supersymmetries"
and "solitons".

The term "catastrophe" was introduced by the French
mathematician René Thom to convey intuitive notions connected
with mathematical schemes to describe such phenomena as dis-
continuities, jumps, corners, surfaces separating homogeneous
phases, biological speciation, etc. Their usefulness in
models of natural phenomena remains under question at the
present time, and has even become a subject of heated debate
in the popular press. The historian of science has the good
fortune of being able to observe an attempt to establish a
new paradigm (in the sense of Thomas Kuhn), and to reflect
upon the social aspects of the process of establishing
scientific public opinion.

"Supersymmetries", studied in supergeometry, are entering
science with less fanfare, although they will perhaps have
greater impact on the development of physics and geometry.

Formally speaking, supergeometry considers not only ordinary functions on a space, say R^n, but also anticommuting functions, i.e., those satisfying the condition $fg = -gf$, from which it follows, in particular, that $f^2 = 0$. Since there are no nonzero numbers with square zero, such a function cannot take numerical values; its natural ranges of values are the so-called Grassmann algebras, introduced in the last century by the remarkable mathematician and Sanskritologist Grassmann. In physics, Grassmann algebras appeared only after the emergence of quantum theory and the concept of spin; it turned out that an adequate description of a collection of identical particles with half-integer spin, such as electrons, requires the introduction of anticommuting variables.

The notion of a soliton arose as a result of the discovery of certain special solutions of equations describing waves in various media--water, for example. The classical wave equations are linear, i.e., the sum of solutions and the product of a solution by a number are also solutions. In other words, these equations describe waves which do not interact with one another. If we want to take into account such properties of real media as dispersion (a nonlinear relation between the frequency and length of an elementary wave) and the nonlinear dependence of the velocity of a wave on its amplitude, we arrive at a much more complicated picture of the interaction of wave perturbations than simple superposition. Therefore, considerable excitement greeted the discovery in the 1960's by a group of American physicists and mathematicians of the effect of a nonlinear addition of certain solitary excitations described by the Korteweg-de Vries equation (these solitary excitations were what was first called solitons).

The height of a soliton wave is proportional to its velocity. Hence, one can attempt to trace the fate of the sum of two solitons that are far apart at an initial moment, where the larger soliton moves toward the smaller one and therefore overtakes it. The common expectation would be that after "collision" the wave picture collapses, but machine experiment shows that nothing of the sort occurs: after a period of interaction the larger soliton "passes through the smaller one", and both begin to diverge, each retaining its form. An exact mathematical theory was constructed shortly after this discovery--it corroborated the preservation of the individuality of the solitons after the interaction, no matter how many there were at the beginning.

Since that time, the number of nonlinear wave equations found to exhibit analogous properties has grown linearly with time, and the number of publications devoted to them has grown exponentially. The hope has been expressed that soliton-like excitations provide an adequate classical image of elementary particles: on a new intellectual level this revitalizes the idea proposed a century ago by Rankine and Thomson that atoms are "vortical rings of the fundamental fluid". The point is that, in contrast to Maxwell's equations, the equations for Yang-Mills connections are nonlinear.

Now a small historical note on the discoverers of the soliton, which the reader may find edifying. (Translator's note: See also a paper by F. van der Blij, "Some details of the history of the Korteweg-de Vries equation", Nieuw Archief voor Wiskunde (3), 26 (1978), 54-64 for some further interesting sidelights.) Diederik Johannes Korteweg was born in Holland in 1848 and died there in 1941. He was a well-known scientist, and several obituaries were dedicated to his memory. None of the obituaries even mentions the now famous paper in which the soliton was discovered. That paper was in essence a fragment of the dissertation of Gustav de Vries, completed under the guidance of Korteweg and defended on December 1, 1894. de Vries was a high school teacher, and almost nothing is known about him.

Sets, Formulas, and the Divided Brain

What is the relation between a mathematical text and its content in the wide sense of the word (the multiplicity of its possible interpretations)? We have attempted to show that an intermediate set-theoretic image must be constructed between, say, Maxwell's equations and their meaning in terms of physical concepts; this construct is an interpretative middleman, functionally similar to the use of an intermediate artificial language in modern linguistic models of machine translation. In fact, a careful analysis of scientific thought would enable one to discover a whole hierarchy of language intermediaries which take part in the potential explanation of such concepts as "number", "photon" or "time".

However, most of these explanations exist in a vague, obscure and unsteady form, frequently accessible to individual cognition, but yielding to communication only to the extent that the means of our natural language can be used. Natural language plays an immense role in the discovery, discussion, and preservation of scientific knowledge, but it is very poorly adapted to the exact transmission of the content of this knowledge, and especially to its processing, which forms an important part of scientific thought. Natural language performs different functions, and ascribes to other values.

The language of modern set-theoretic mathematics can affect the role of such a language-intermediary thanks to its unique capacity to form geometric, spatial and kinematic forms, and at the same time to provide a precise formalism for transcribing their mathematical content. Cantor's definition of a set, which was given above, has with some irony been called "naive" by comparing it with Euclid's definition of a point as a "place with neither length nor breadth". This criticism is testimony to the lack of understanding of the fact that the fundamental concepts of mathematics, which are not reducible in a given system to more elementary concepts, must necessarily be introduced in two ways: the concrete ("naive") way and the formal way. The purpose of the concrete definition is to create a prototypical, not yet completely formed shape, to tune different individual intellects to one scale, as by a tuning fork. But the formal definition does not really introduce a concept, but rather a term; it does not introduce the idea of a "set" into the structure of the mind, but rather introduces the word "set" into the structure of admissible texts about sets. These texts are described by rules for generating them, in much the same way as instructions in ALGOL describe rules for composing programs. In the limit of idealization, all of mathematics can be regarded as the collection of grammatically correct potential texts in a formal language.

In this image there is a strange, and to many appealing, esthetics of ugliness. It arose in the works of thinkers who thought deeply about how to reconcile belief in the absolute truth of mathematical principles with the abstractions of infinite sets and infinite processes of verification used to introduce this truth. The original conjecture of David Hilbert was that, strictly speaking, these abstractions do not need any "almost physical" and thus questionable inter- pretation, and that they can be taken to be purely linguistic facts. "Infinity" is not a phenomenon--it is only a word which enables us somehow to learn truths about finite things. We have already mentioned that Gödel later showed that such a linguistic notion of "provably true" is far narrower than the abstraction of truth that can be developed through ideas of infinite verifications.

The external, scientific and applied (in the broad sense of the word) aspects of mathematical knowledge, when analyzed epistemologically, allow us to understand something about mathematical creation and the dialectics of its interrelation with Gödel's prohibition. The adoption of an interpretation of formalism which is in some sense physical, the belief in the adequacy of this interpretation and the knowledge of certain features of the behavior of a physical phenomenon, allow us to guess or to postulate mathematical truths that are not accessible to "pure intuition". This is the source for extending the very basis of mathematical knowledge.

On a more special plane, it has become possible in recent years to consider the relation between mathematical symbolism, informal thought and the perception of nature from a point of view provided by new data about the structure and functions of the central nervous system.

The brain consists of two hemispheres, a left hemisphere and a right one, which are cross-connected with the right and left halves of the body. The neuron connections between the two hemispheres pass through the corpus callosum and commissures. In neurosurgical practice there is a method of treatment of severe epileptic seizures, in particular, which consists of severing the corpus callosum and the commissures, thereby breaking the direct connections between the hemispheres. After such an operation patients are observed to have "two perceptions". In the laconic formulation of the American neuropsychologist K. Pribram, the results of studies of such patients, and also patients with various injuries of the left and right hemispheres, can be summarized in the following way: "In right-handed persons the left hemisphere processes information much as does the digital computer, while the right hemisphere functions more according to the principles of optical, holographic processing systems." In particular, the left hemisphere contains genetically predetermined mechanisms for understanding natural language and, more generally, symbolism, logic, the Latin "ratio"; the right hemisphere controls forms, gestalt-perception, intuition. Normally, the functioning of human consciousness continually displays a combination of these two components, one of which may be manifested more noticeably than the other. The discovery of their physiological basis sheds light on the nature and typology of mathematical intellects and even schools working on the foundations of mathematics. One might conjecture that the two great intellects standing at the cradle of modern mathematics, Newton and Leibniz, were, respectively, right-hemispherical and left-hemispherical types. To Newton we owe the conception of mathematical analysis and the first fundamental results of mathematical physics: the universal law of gravitation, and the derivation of Kepler's laws and the theory of tides from it. On the other hand, Leibniz invented calculus and introduced notation, in particular $\int y\,dx$, which we still use today. In the words of the historian of mathematics D.J. Struik, he was "one of the most fruitful inventors of mathematical symbols", and even his version of mathematical analysis was invented as a result of a search for a universal language. (It is interesting that Newton, who also did not escape this craze of his time, nevertheless did not invent the formalism of analysis and presented his proofs geometrically.)

If we accept modern notions concerning the functional asymmetry of the brain, we can conjecture that set theory allows the working mathematician to achieve by the shortest path a properly balanced activity of the left and right hemispheres of his brain; it is this which explains its remarkable effectiveness.

In conclusion, I would like to cite the words of I.A. Sokolyansky, who devoted his life to the education of blind and deaf children. The quote is contained in a letter to Vyacheslav V. Ivanov, from whose book "Even and Odd, Asymmetry of the Brain and Symbolic Systems" (Moscow, "Soviet Radio", 1978) the following information is taken.

If those parts of the central nervous system of a blind and deaf child that control visual perception of the outside world are not damaged, then the child can be taught a language, even speech, and the complete development of its personality can be ensured. This process is accomplished in several steps. First, the child is supported by constant contact with the mother or a teacher, holds onto her hand or skirt, walks around the house, touches and feels the objects of her actions, and on this basis manufactures a language of gestures which imitates to some extent the actions and properties of objects. Normally this is a function of the right hemisphere. Sokolyansky's discovery was that in the next stage the child can be taught to recode the language of gestures into a finger alphabet, so that a gesture-hieroglyph is replaced by a gesture-word. The symbol of translation is a special gesture similar to the mathematical equal sign--two palms extended parallel. The meaning of this recoding is that the information is transmitted to the left hemisphere, which since it is predisposed to be taught discrete and

symbolic language (not necessarily spoken language!), begins
to develop this function with practically the same speed as
in a normal child--mastery of the language takes two to three
years. The syntax of this left-hemisphere language is
different from the "real-world" syntax imbedded in the right
hemisphere, and is more or less identical to the syntax of
verbal natural language. The semantics, however, is apparently more restricted, or in any case not quite equivalent to
the semantics of someone who sees and hears. How does one
explain what "star" means to someone who has never seen a
star? Sokolyansky gives a noteworthy answer: "Verbal speech,
although it can be mastered by the dumb, cannot by itself
guarantee the deaf-blind complete intellectual development to
the extent of reflecting the external physical world as it is
accessible to a normal person. The true picture of this
world can be uncovered only by a mathematically developed
intellect..."

Even those who see stars ask "What is a star?", because
to see merely with one's eyes is still very little.

2. PHYSICAL QUANTITIES, DIMENSIONS AND CONSTANTS:
The Source of Numbers in Physics

> "The whole purpose of physics is to find a number, with
> decimal points, etc.! Otherwise you haven't done
> anything." (R. Feynman)

This is an overstatement. The principal aim of physical
theories is understanding. A theory's ability to find a
number is merely a useful criterion for a correct understanding.

Numbers in physics are most often the values of physical
quantities that describe the states of physical systems.
"Quantities" is the generic name for such abstractions as
distance, time, energy, action, probability, charge, etc. A
state of a system is characterized, in turn, by the values of
a sufficiently complete collection of physical quantities;
and it is most natural to describe a system by giving the set
of its possible states. One cannot escape from this vicious
circle if one restricts oneself to purely verbal descriptions.
The circle can be broken in two ways--operationally, when we
explain how to measure the mass of the Earth or an electron,
and mathematically, when we propose a theoretical model for a
system or a class of systems and assert that the mass m is,
say, the coefficient in Newton's formula $F = ma$.

Substantive, although simple, mathematics connected with physical quantities begins with a reminder that the values of a physical quantity (more precisely, a real scalar quantity) can be identified with numbers *only after a choice of unit of measurement and origin of reference (zero point)*. It is reasonable to refrain from introducing this element of arbitrariness for as long as possible. In fact, some of the most fundamental physical laws state that certain physical quantities have natural units. We will examine this in more detail.

The Spectrum of a Scalar Quantity

The spectrum of a quantity is the set of all the values that the quantity can take (on the states of a given system, of a given class of systems, or of "all" systems--this is made more precise as needed). The basic mathematical postulate that can be considered as the definition of a scalar quantity in theoretical models is that the spectrum is always a subset of one-dimensional affine space over the real numbers. In other words, it lies on a line where zero and one are not marked; if two points are specified to be zero and one, then the spectrum becomes the set of real numbers. The gist is that sometimes these points need not be distinguished arbitrarily, but rather one can use the spectrum itself. Here are some basic examples.

(a) Velocity. The smallest (relative) velocity is naturally called "zero". The second distinguished point on the spectrum of velocities is c, the speed of light. The conventional assumption about the spectrum of velocities (since the creation of the special theory of relativity) is that it fills out the interval from zero to c. Then it is natural to declare c to be the unit of velocity and to take all velocities in the segment [0,1]; in the more usual notation, this is true of the ratio v/c. In everyday life we rarely encounter velocities greater than 10^{-6} on this scale (the speed of sound).

(b) Action. This is perhaps the most important quantity of all in theoretical physics, and we shall devote a separate chapter to it. It takes values on intervals of the history, or paths, of a physical system, not on instantaneous states. In classical physics it determines the dynamics, i.e., the physically possible path intervals on which "action" takes the smallest admissible values. The natural zero on the spectrum of actions is the action of the "infinitesimal" history of the system. The upper boundary of the action spectrum is unknown. It may exist in a cosmological model, where this boundary will be the action of the Universe in its entire history from the Big Bang to the Big Collapse, if the latter is predicted by the model.

Nevertheless, another distinguished point is known on the action spectrum: this is the famous Planck constant h. In human terms it is extremely small--the action of a pen writing the word "action" is of the order of 10^{29} to 10^{30}h. Pragmatically speaking, h indicates when quantum-mechanical models must be used rather than classical ones: namely, in cases when we are interested in details of the history of the system where the action only varies over a few h. (However,

this condition is neither necessary nor sufficient.)
Choosing h as a unit of action, we may take the action
spectrum to be the half-line $[0, \infty)$, and the spectrum of in-
crements of action to be the whole real line.

Thus, the point h is "unseen" on the action spectrum,
in contrast, say, to the speed of light c, which is the
right endpoint of the velocity spectrum. This is very
strange. However, there are two contexts in which h appears
explicitly.

One of these is connected with the spin spectrum--the
internal angular momentum of elementary particles. Spin has
the same dimension as action, and consists of integer mul-
tiples of $\hbar/2 = h/4\pi$. Does this not mean that spin is truly
the fundamental quantity, and action only a vestige of classical physics?

The second context is the famous Heisenberg uncertainty
principle. Quantum models define a partition of the classical
system of quantities into conjugate pairs: (coordinate of
position, projection of momentum), (energy, time). A product
of conjugate quantities has the dimension of action. In a
verbal formulation the uncertainty principle asserts that
both members of a pair of conjugate quantities cannot simultaneously take an exact value on any state of the system.
The product of the imprecisions is bounded from below by $\hbar/2$.
Applying this principle to energy and time, we formally
obtain the relation $\Delta E \cdot \Delta t > \hbar/2$, whose substantive meaning
has been discussed many times in the physical literature.
From our point of view this means that the notion of a

classical segment of the history of the system on which the action varies by less than $\hbar/2$ is devoid of meaning. Later on, we discuss in more detail the difficult question of the comparative meaning of classical and quantum quantities having the same name.

(c) Mass. According to Newton, values of inert mass can be ascribed to stable material bodies. The smallest objects to which the Newtonian concept of mass is still applicable without fundamental reservations are the electron and the proton. They give us two points on the spectrum of masses (other than zero): m_e and m_p. A typical mass on the human scale can be given using Avogadro's number $6.02 \times 10^{23} \, m_p$. The ratio $m_p/m_e \approx 1840$ is the first truly fundamental number that we have encountered so far, in contrast to the points of the spectrum, which are not, strictly speaking, numbers.

Other elementary particles determine other points on the spectrum of masses; measuring them in units of m_e or m_p, we obtain a bunch of numbers for which theoretical explanations are needed.

(d) Gravitational constant   If two point masses m_1 and m_2 are found at a distance r from each other and are attracted to each other with a force F resulting only from their Newtonian gravitational attraction, then the quantity $Fr^2/m_1 m_2$ does not depend on m_1, m_2, or r. This quantity was discovered by Newton and is denoted G.

This last example is conceptually more complicated than the preceding ones: to introduce G we must explicitly appeal to a "physical law". Moreover, we have obtained a point of a new spectrum--the spectrum of constants of fundamental interactions, which also include electromagnetic, strong and weak interactions.

It is appropriate at this point to introduce the following large group of physical abstractions.

Physical Law, Dimension and Similarity

For the purposes of this section "physical law" means the content of such formulas as $F = ma$, $F = Gm_1m_2/r^2$ (Newton), $E = h\nu$ (Planck), $E = mc^2$ (Einstein), etc. A physical theory, such as Newtonian mechanics or Maxwell's electromagnetic theory, includes the following data on the mathematical side: (a) the basic quantities of the theory, and (b) the basic laws connecting them. Moreover, on the operational side a theory must describe: (c) the physical situations in which the theory can be applied, and (d) principles for comparing the theoretical statements with measurements and observations.

Here we are concerned only with the first part. Knowing the quantities and the laws connecting them, we can construct a fundamental mathematical characteristic of a theory--its group D of dimensions. In mathematical language this is an abelian group which can be defined by generators and relations: the generators are the physical quantities of the theory, and the relations are determined by the condition

that all the laws of the theory are homogeneous. The class
of a quantity in the group D is called the dimension of
this quantity. It is possible to choose basic quantities
which form an independent system of generators in the group
D; the dimensions of the remaining quantities of the theory
can be expressed in the form of formal monomials in the
basic ones. The units of the basic quantities determine the
units of the others. (We are giving a condensed presentation of the principles underlying such notation as cm/sec^2.)
We note several circumstances in which the explicit introduction of the group D helps to elucidate the essence of
the theory.

The Group of Dimensions of Newtonian Mechanics

It is generated by the dimensions of length L, time T
and mass M. The law $F = ma$ shows that force has dimension
MLT^{-2} in this group, while energy (force times length) has
dimension ML^2T^{-2}, and action (energy times time) has dimension ML^2T^{-1}.

Progress in physics is continually accompanied by two
opposing processes: *enlargement* of the group D of dimensions through the discovery of new kinds of quantities
(electromagnetism after Newton; new quantum numbers such as
"strangeness" and "charm" in our days), and the *shrinking* of
the group through the discovery of new laws giving relations
between previously independent dimensions.

In order to understand this second process we return to the Newtonian gravitational constant G. According to our rules, its dimension is force $\times$ (length)^2 $\times$ (mass)$^{-2}$ $= M^{-1}L^3T^{-2}$. Thus, its numerical value depends on choosing units of mass, length and time. It is a constant in the sense that, once we choose these units, the numerical value we obtain using the formula Fr^2/m_1m_2, where F, r, m_1, m_2 vary in different Eötvös type experiments or are calculated from astronomical data, does not depend on the variable quantities in these experiments: r, m_1, m_2.

After establishing this physical fact, we can use it to construct a smaller group D' of dimensions for the theory "Newtonian mechanics + Newtonian gravitation". Mathematically, this smaller group is the quotient group of D by the subgroup consisting of all powers of $M^{-1}L^3T^{-2}$. For the basic dimensions in D' one can take any pair (M,L), (M,T) or (L,T); the remaining dimensions can then be expressed in terms of this pair and the dimension G. The number of fundamental units for D' then decreases to two if we take G as a unit of measurement of dimension $M^{-1}L^3T^{-2}$. From this example we also see the physical meaning of the distinguished points of spectra: they are points that can be reproduced in a series of experiments of a certain type which isolate certain interactions, certain types of systems, and so on.

Scale Invariance

The group D* of similarities, or scale invariances, of a given theory can be defined from the mathematical point of

view as consisting of the *characters* of the group D of dimensions, i.e., the mappings χ from the group D to the positive real numbers with the multiplicativity property:
$\chi(d_1 d_2) = \chi(d_1)\chi(d_2)$ for all d_1, $d_2 \in D$. This group has a
direct physical meaning: it shows the proportion by which
one can increase (or decrease) the various characteristics of
a phenomenon before it leaves the range of applicability of
the theory. If we know all the laws of the theory, then D*
is trivial to calculate. The use of D* is that sometimes
it can be guessed from physical considerations before the
exact form of these laws becomes known. Then it turns out
that D* carries important information about these laws. A
well-known example of a classic discovery made in this way is
Wien's law $\varepsilon(\nu,T) = \nu^3 F(\nu/T)$ for the radiative capacitance of
an absolutely black body as a function of frequency and temperature. It corresponds to the character $\chi_a([\nu]) = a$,
$\chi_a([\varepsilon]) = a^3$, $\chi_a([T]) = a$ in the group D*, where $[\varepsilon]$, $[\nu]$,
$[T]$ are the dimensions and a is an arbitrary real number.
One can also mention Galileo's arguments on the sizes of
living creatures and the numerous applications of the theory
of similarity in hydro- and aerodynamical calculations. On
the level of fundamental theories the group D* is the
simplest example of a symmetry group. In elementary particle
physics and quantum field theory such groups are more and more
frequently considered as independent physical laws on a higher
level, imposing rigid restrictions on laws on the next level,
such as Lagrangians. The most important of these groups are
noncommutative and complex, such as groups of unitary

rotations in U(n), because in quantum mechanics the basic quantities lie in higher-dimensional complex spaces, and not in one-dimensional real ones. But that is already another story.

Planck Units and the Problem of a Unified Physical Theory

In Newtonian physics there are no other natural units besides G. The speed of light c can be declared a natural unit only within a new theory which postulates its particular role as the upper limit of the speed of propagation of material bodies (unattainable) or signals (attainable), as an invariant relative to a change of inertial system of coordinates, etc. In a similar way, Planck's unit of action $\hbar$ became the coat of arms of a new physical theory--quantum mechanics.

In the Newtonian group D, both c and $\hbar$, like G, have completely determined dimensions: LT^{-1} for c and ML^2T^{-1} for $\hbar$. Having chosen G, c and $\hbar$ as the basic units for the corresponding dimensions, we discover that there is a natural scale which makes real numbers out of the values of all classical physical quantitites expressible in D; i.e., there are natural units for all quantities! In fact, the dimensions $M^{-1}L^3T^{-2}$ (of G), LT^{-1} (of c) and ML^2T^{-1} of $\hbar$) generate the whole group D (to be really precise, they generate a subgroup of index two).

In particular, the natural units of length, time and mass--the celebrated units of Planck--are:

$$L* = (\hbar G/c^3)^{1/2} = 1.616 \times 10^{-33} \text{ cm};$$

$$T* = (\hbar G/c^5)^{1/2} = 5.391 \times 10^{-44} \text{ sec};$$

$$M* = (\hbar c/G)^{1/2} = 2.177 \times 10^{-5} \text{ g}.$$

The reader might ask: Why don't we ever measure in Planck units? The pragmatic answer: because they give a completely absurd scale. The Bohr radius is equal to 3.9×10^{-11} cm: $L*$ is 22 orders of magnitude less! $T*$ is the same number of orders of magnitude less than the time it takes light to pass through a Bohr radius. On the other hand, $M*$ is the mass of a real macroscopic particle containing about 10^{19} protons. The Planck unit of density $M*/L*^3$ is equal to 5×10^{93} g/cm^3; we cannot imagine anything having density remotely similar to this under any circumstances.

A more substantive remark is that, in fact, we have no unified physical theory in which G, c and $\hbar$ occur simultaneously. Even the theories which combine these constants in pairs are considered the greatest achievements of the twentieth century: (G,c) is general relativity theory; (c,$\hbar$) is relativistic quantum field theory, which is in relatively complete form only for electromagnetic interactions. The estimates of the quantum generation of particles in strong classical gravitational fields, in particular, close to black holes of small mass (S. Hawking), are related to the fragments of a future (G,c,$\hbar$)-theory. As long as a complete quantum (G,c,$\hbar$)-theory does not exist, Planck units remain the remote outposts of an extensive uncharted territory.

There is another point of view regarding this strange disparity of orders of magnitude of natural units with one another and with more common units. In the hypothetical fundamental equations of a unified theory--"the universal theory of everything" (Stanisław Lem)--the different terms (now dimensionless numbers!) will, thanks to this disparity, take values that differ very sharply in magnitude depending on the scales of the domain of space/time, momentum/energy in which the phenomena to be studied are located. Very small terms can be discarded with negligible error, leading to one of the approximate theories. This also happens at the limits of applicability of the currently known models, when we describe the world classically in human scales (supposing $v/c = 0$ and $\hbar/S = 0$, where v is a typical velocity and S is a typical action), or when we neglect gravitation in micro-scales, setting $G = 0$.

In fact, this argument, though it contains a grain of truth, is very naive. A real change of theory is not a change of equations--it is a change of mathematical structure, and only fragments of competing theories, often not very important ones conceptually, admit comparison with each other within a limited range of phenomena. The "gravitational potential" of Newton and the "curvature of the Einstein metric" describe different worlds in different languages.

Moreover, life--perhaps the most interesting physical phenomenon--is embroidered on a delicate quilt made up of an interplay of instabilities, where a few quanta of action can

have great informational value, and neglect of the small
terms in equations means death.

Classification of Physical Constants

We sum up what has been said. The reference book *CRC
Handbook of Chemistry and Physics* contains 2431 pages of text and
many millions of numbers. How can one understand them?
These quantities can be divided into at least four types.

(a) Natural units of measurement , or physically distinguished
points of spectra. These are not numbers but entities such as
G, c, h, m_e, e (charge of an electron). These are dimensional characteristics of certain phenomena which are capable
of being reproduced again and again with a high degree of
accuracy. This is a reflection of the fact that nature produces elementary situations in large quantities. Contem-
plating the identical nature of each type of building block
of the universe has from time to time led to such profound
physical ideas as Bose-Einstein statistics and Fermi-Dirac
statistics. Wheeler's fantastic idea that all electrons are
identical because they are the instantaneous cross-sections
of the tangled-up world-line of a single electron, led Feynman
to an elegant simplification of the diagrammatic computation
technique in quantum field theory.

(b) True, or dimensionless, constants. These are ratios of
various distinguished points on the spectrum of a scalar
quantity, for example, the ratios of masses of electrical
particles (we have already mentioned m_p/m_e). The identifi-
cation of different dimensions because of a new law, i.e.,

the reduction of the group of dimensions, has the effect of combining once-distinct spectra and giving birth to new numbers which must be explained.

For example, the dimensions m_e, c and $\hbar$ generate the Newtonian group and therefore lead to atomic units for the dimensions M, L and T which are as natural as the Planck units. Thus, their ratios to the Planck units require a theoretical explanation. But, as we have said, this is impossible as long as there is no $(G, c, \hbar)$-theory. However, even in the $(m_e, c, \hbar)$-theory--quantum electrodynamics--there is a dimensionless constant to which modern quantum electrodynamics in some sense owes its existence. We place two electrons at a distance $\hbar/m_e c$ (the so-called Compton wavelength of an electron) and measure the ratio of the energy of their electrostatic repulsion to the energy $m_e c^2$, which is equivalent to the rest mass of an electron. We obtain the number $\alpha = 7.2972 \times 10^{-3} \approx 1/137$. This is the famous fine structure constant.

Quantum electrodynamics describes, in particular, processes which do not preserve the number of particles: a vacuum generates an electron-positron pair, which are annihilated. Because the energy of generation (not less than $2m_e c^2$) is 100 times greater than the energy of the characteristic Coulomb interaction (thanks to the value of α), one can carry out an effective scheme of calculations in which these radiation corrections are not completely neglected, but do not hopelessly "spoil the life" of the theoretician.

There exists no theoretical explanation for the value of α.

Mathematicians have their own remarkable spectra: the spectra of various linear operators (for example, the generators of simple Lie groups in irreducible representations), the volumes of fundamental domains, the dimensions of homology and cohomology spaces, etc. There is wide scope for fantasy in identifying the spectra of mathematicians and the spectra of physicists. What we need instead is principles to limit the choices. Let us return, however, to constants.

The next type occupies a large part of tables of physical quantities:

(c) Coefficients for converting from one scale to another, for example, from atomic scales to "human" scales. Among these constants are: the Avogadro number mentioned before, $N_0 = 6.02 \times 10^{23}$, essentially one gram expressed in "proton masses", although the traditional definition is a bit different; and also such things as a light year expressed in kilometers. For mathematicians the most disagreeable aspect of this is, of course, that the coefficients convert from one set of physically meaningless units to another which is equally meaningless: from cubits to feet or from degrees Réaumur to degrees Fahrenheit. In human terms these are sometimes very essential numbers; as Winnie-the-Pooh remarked philosophically: "But whatever his weight in pounds, shillings and ounces, He always seems bigger because of his bounces."

(d) "Diffuse spectra " These characterize materials (not elements or pure compounds, but ordinary industrial brands of steel, aluminum, copper), astronomical data (the mass of the Sun, the diameter of the Milky Way...) and many other such things. In contrast to the situation with electrons, nature produces stones, planets, stars and galaxies with no concern for creating exact similarity; but, nevertheless, their characteristics vary only within rather definite limits. The theoretical explanations for these "permitted zones", when they are known, tend to be quite interesting and instructive.

V. Weisskopf gathered together a series of such explanations in his excellent article "Contemporary physics in an elementary presentation" (CERN Report No. 70-8). Here is an example of a physical argument from this paper, in which all of our principal heroes play their parts: "heights of the mountains in terms of fundamental constants". Take the following question: the Earth's highest mountain Chomolungma (Everest) has a height of about 10 km; why are there no higher mountains? It turns out that, even without taking into consideration the geological mechanisms of erosion and weathering, we find that the height of a mountain is limited by a few tens of kilometers because of the concrete dimensions of the Earth and the values of the fundamental constants. Weisskopf argues as follows: a mountain of too great a height cannot exist because of the liquefaction of its lower part under the pressure of the upper part. A calculation of the height at which the pressure is almost sufficient for liquefaction yields the estimate:

$$\gamma \, \frac{\alpha a_0}{\alpha_G} \cdot \frac{1}{N^{1/3}} \cdot \frac{1}{A^{5/3}} \approx 40 \text{ km} \quad ,$$

where $\gamma = 0.02$ is the melting heat characteristic (which can be estimated entirely in terms of the fundamental constants); α is the fine structure constant, $\alpha_G = Gm_p^2/\hbar c$; $N \approx 3 \times 10^{51}$ is the number of protons and neutrons in the Earth; $A \approx 60$ is the average atomic weight of the matter in a mountain. Only the number N here is not fundamental. But even its place on the diffuse spectrum of the planet masses is bounded by the fundamental constants. Using very crude estimates, Weisskopf shows that N could not exceed about 10^{53}, since otherwise the matter in the planet could not exist as un-ionized atoms. Finally, one can estimate N from below by requiring that the height of mountains on a planet be no greater than the radius of the planet, i.e., that the planet be approximately spherical, since otherwise mountains cannot be distinguished as such! This estimate leads to the size of large asteroids.

3. A DROP OF MILK:
Observer, Observation, Observable and Unobservable

"...What would be observed (if not with one's actual
eyes at least with those of the mind) if an eagle,
carried by the force of the wind, were to drop a rock
from its talons?"   (G. Galilei)

The actual eyes of my contemporaries observed how a bomb
flies when the locks of a bomber open against a smoky sky,
on the newsreel screens and in thousands of children's draw-
ings; I myself drew them.   Let us try to forget this and look
at the world with the eyes of the mind, as our immortal contemporary Galileo Galilei taught the simple-minded Simplicio.

Isolated Systems

Among all the abstractions of classical physics one of
the main ones is the idea of an isolated, or closed, system.
This is a part of the Universe whose evolution over some
period of its existence is determined only by intrinsic laws.
The external world either does not interact with the system
at all, or in some models this interaction is incorporated
summarily, with the help of such notions as constraints, an
external field, a thermostat, etc. (thus we have used the
words "isolated" and "closed" more broadly than is usually
done; the isolation refers, rather, to the mathematical
model).   There is no feedback loop, or if there is, it is

artificially severed. The world is divided into parts, modules and assemblies, as in industrial specifications. And in fact this is the ideology not only of *Homo Sapiens* but of *Homo Faber*. The screws and gears of the great machine of the world, when their behavior is understood, can be assembled and joined in a new order. Thus one obtains a bow, a loom or an integrated circuit.

For the mathematician an isolated system consists of: (a) its phase space, i.e., the set of possible instantaneous states of motion of the system; (b) the set of curves in phase space describing all possible histories of the system, i.e., sequences of states through which the system passes in the course of time. The first of these data is kinematics, the second is dynamics. It is important to distinguish a state of the system from a state of its motion: the first is traditionally given by coordinates, and the second by coordinates and velocities. Knowing only the coordinates, we cannot predict the later motion of the system, but we can if we know the coordinates and the velocities. The assumption that a closed system can be described by a convenient phase space and a system of curves in it (the whole thing together is often called a phase portrait of the system) is the mathematical content of the classical principle of determinism.

One of the famous paradoxes of Zeno of Elea can be interpreted as a first approximation to understanding the role of the phase space: a flying arrow and a stationary arrow at each moment of time are located at the place where they are located; what then distinguishes flight from immobility?

Answer: the apparent place of the arrow is only the projection onto the space of positions of its "true place" in
the space of pairs (position, velocity vector).

A classical closed system is isolated from the whole
external world; thus also from the external observer, and
from the effect which an observer can have on it. Observation is not interaction. Observation is the most important
thought-experiment which can be performed on the system, and
its first goal is to localize the system in its phase space.
The converse can also be said: the phase space is the set of
all possible results of complete instantaneous observations.
A complete observation allows one to compute the full evolution of a classical system; the existence of complete observations is another form of the determinism postulate. Evolution is the set of results of observations at all moments of
time. The possibility of a mental observation with no interaction is confirmed by considering different methods of
observations which approximate reality and for which the
interaction is part of the design, but can be made as small
as desired or completely accounted for in computations, i.e.,
it is controllable. In essence, one considers an isolated system
to be included as part of a larger system (S,T). An observation corresponds to an act of weak interaction between S
and T which scarcely interrupts the evolution of S (perhaps the interaction is switched on for a short time and then
cut off). It is of principal importance that the abstraction
of conceptual observation can then be applied to the combination (S,T) without influencing the evolution of the

combined system. Moreover, we suppose that S can become part of (S,T) for a short time without losing its individuality, i.e., it can do so reversibly.

This is a very natural postulate for man, whose principal means of observation is vision. Electromagnetic interactions are so weak that in all scales from the cosmic to the human a glance at the system does not act on it at all.

The "eyes of the mind" must be able to see in the phase space of mechanics, in the space of elementary events of probability theory, in the curved four-dimensional space-time of general relativity, in the complex infinite dimensional projective space of quantum theory. To comprehend what is visible to the "actual eyes", we must understand that it is only the projection of an infinite dimensional world on the retina. The image of Plato's cave seems to me the best metaphor for the structure of modern scientific knowledge: we actually see only the shadows (a shadow is the best metaphor for a projection).

For a human it is psychologically very difficult to transcend the limits of the usual three spatial dimensions. But we only mislead ourselves if we awkwardly try to describe the internal quantum degrees of freedom as the "value of the projection of spin on the z-axis", since a spin vector lies in a completely different space from the z-axis. It is worth remembering that even the three-dimensionality of the world was recognized after great effort--it was taught to us by the artists of the Renaissance. Uccello withdrew from his affairs for ten years in order to devote himself to the study of perspective. Among other things, modern mathematics is rigorous training in multidimensional perspective, following a unified program. If we believe the neuropsychologists, in the course of this training the left and right halves of the brain behave as the blind man and the legless guide he carries on his back.

Classically, an observer is represented, in general, by a system of coordinates in the fundamental spaces of the theory. A unit of measurement determines a coordinate in the spectrum of the quantity being measured. When these units are chosen, the coordinate functions, i.e., the observables, identify the space of positions, the phase space or parts of them with subsets of the R^n and C^n of mathematicians. A theory with observables is good in that it simultaneously describes both the concepts and their observable "shadows". A theory with observables is bad in that it may turn out to be simpler, more instructive and more straightforward to explicitly separate as soon as possible the observed from the observer and investigate their interrelation as a new object of study.

According to Newton and Euler, color is the spectral composition of light radiation in the band of wavelengths around half a micron; according to Goethe, color is what we see. It is striking to what extent these two notions do not lend themselves to a direct comparison--they are connected only by a complicated physiological theory of color vision. The humanist Goethe could not allow a denial of the observer, since his entire system of values could not exist without the idea of human participation as a measure of things.

Much can be said for this point of view. Much can also
be said against it: often the best method to find oneself is
to turn away from oneself. The Newtonian theory of light--
and all later physical theories--were called upon to explain
what light is without any relation to the fact that it can be
seen. For Goethe the principal thing about light is that it
can be seen. And again it turns out, as always, that the
visible must be explained in terms of the invisible.

"All motions that a celestial body is observed to under-
go belong not to itself, but to the Earth" (Copernicus, 1515).
After this phrase, which moved the Earth, all subsequent
theories based only on the notion of "observable" became
archaic before they were born.

The classical observer lives in a world of human scales,
and the conception of a classical observer undergoes natural
changes when one passes to cosmological or microscopic scales.
Distance, time, energy and motion are so great for astronomical phenomena that the hypothesis that the observer does not
influence events seems acceptable without further discussion.
Other problems of observation appear in the foreground. Two
of them can be summarized briefly as questions. Can the
Universe be considered as a closed system? What should be
our attitude toward a theory describing phenomena that cannot
be observed because of their destructive impact on the observer or because some regions of space-time are fundamentally
isolated from observation (stellar temperatures, masses,

pressures, gravitational fields of black holes, conditions of the Big Bang)?

The very principles of describing closed systems are based on the hypothesis that such systems can be reproduced-- the phase space of a system realizes the idea of the feasibility of different states and different paths of evolution. How does one reconcile this idea with the uniqueness of the evolution conveyed to us in observations of the system? The answer, of course, lies in the notion of the local interaction of "parts of the world" with one another, and the essential sameness of the laws of physics that act in different parts. In the simplest and most fundamental models of the Universe (the Friedmann model, the Einstein-de Sitter model), the idea of homogeneity occurs, manifesting itself in the existence of a large group of symmetries of the mathematical model.

In all models of cosmology one finds in the foreground the idea of modeling a closed system. In describing an isolated system, we not only imagine that it does not interact with the external world; we also oversimplify its internal structure, ignoring many "insignificant details". Now when the World is this "isolated system", then everything even remotely associated with our daily life shrinks into the ranks of "insignificant details". In a cosmological model of the World not the slightest trace of the familiar remains.

But in spite of this, or perhaps thanks to it, a paper in "Uspekhy Fizicheskikh Nauk" may begin with the phrase: "We would be happy if Cygnus X-1 turns out to be a black

hole" (Usp. Fiz. Nauk 126 (1978), no. 3, p. 515). We know
something about the world because we are happy to perceive it.

Principles of Quantum Description

An ideal observer of the macroworld cannot change it,
but even an ideal observer of the microworld is bound to
change it. This is explained in numerous treatises on quantum
mechanics, but it seems to be poorly understood. Quantum
mechanics does not simply teach us new mathematical models of
phenomena; it furnishes a model of a new relation between
description and phenomenon. In particular, several characteristics of these models are explained verbally by drawing
on the idea of "unobservability". The meaning of this word
changes its inflections like Proteus; the following are unobservable: the phase of the psi-function, a virtual photon,
the color of a quark, the difference between identical particles, and much more.

Let us try to look at the geometry of quantum mechanics
with the "eyes of our mind".

Phase Space

The phase space of a closed quantum system is a set of
rays (one-dimensional subspaces) in a complex vector space $\mathcal{H}$
in which a scalar product is also given. In this postulate
we express: (a) the principle of linear superposition;
(b) the principle of "unobservability of the phase". Instead
of using a whole line in $\mathcal{H}$ to describe the state of the
system, we usually consider a vector lying in this line. It

is defined only up to multiplication by a complex number. Even if we normalize the vector by requiring that its length be equal to one, there still remains an arbitrariness in the choice of a factor $e^{i\theta}$. This θ is the "unobservable phase".

Phase Curves

In order to describe them we must explain how each ray in $\mathcal{H}$ varies with time t, as it traces the evolution of the isolated system. The standard description is as follows. (a) In $\mathcal{H}$ there are N mutually orthogonal rays which do not vary at all: they correspond to the stationary states of the system. (Here N is the dimension of $\mathcal{H}$; as in Chapter 1, for simplicity we only consider the finite dimensional case.) (b) To each of the stationary states ψ_j there corresponds a quantity E_j with the dimension of energy, which is called the energy level of the corresponding station- ary state. If at time zero the system is in the state $\Sigma a_j \psi_j$, then at time t it will be in the state $\psi(t) = $ $= \Sigma a_j \psi_j e^{E_j t/i\hbar}$. We note that $E_j t$ has the dimension of action and is naturally measured in Planck units $\hbar$. Since $e^{Et/i\hbar} = \cos(Et/\hbar) - i \sin(Et/\hbar)$, each term here is periodic in time, and essentially describes motion around a circle with an angular velocity. Thus, their sum represents a rotation around the N axes with different velocities. The trajectories of two-dimensional motions of this sort are the well-known Lissajous figures. (Another image from the ancient history of science is the epicycles of Ptolemy, which also lead to a sum of circular motions.) Any coordinate of the vector $\psi(t)$ experiences frequent and extremely irregular oscillations with time; the graph of even the simple function $\sum_{n=1}^{10} \cos(n^2 t)$ looks like a seismogram. It is convenient to write $\psi(t)$ in the form $e^{-iS(t)}\psi(0)$, where $S(t)$ is the linear operator "action during the time interval $[0,t]$".

Observation: Ovens, Filters and Quantum Jumps

The classical idealization of an observer capable of fixing the instantaneous position of a system on its phase curve is replaced by a radically new system of concepts. At first we shall not call them by their commonly accepted names, in order not to create illusions. A highly idealized assumption about the connection of the above-described scheme with reality is that for each state $\psi \in \mathcal{H}$ one can make a physical device (an "oven") A_ψ which produces the system at the state ψ. Moreover, for each state $\chi \in \mathcal{H}$ one can make a device (a "filter") B_χ which systems enter in the state ψ, and which the system when it leaves can either be detected at the state χ or not detected at all ("the system does not pass through the filter").

The third basic postulate of quantum mechanics (after the superposition principle and the law of evolution) is the following: *a system which is prepared in the state ψ and immediately after this passes through the filter* B_χ *will arrive at the state χ with probability equal to the square of the cosine of the angle between the rays ψ and χ in $\mathcal{H}$.* If a time t elapses between the preparation of the system at ψ and its admission through the filter B_χ, then the probability will be equal to the square of the cosine of the angle between $e^{-iS(t)}\psi$ and χ.

This can be interpreted as follows. As long as nothing is done to the system, it moves along its phase curve. But as soon as it hits a filter which produces only systems at the state χ, its state vector changes by a jump--either it rotates by the angle between ψ and χ, and the system passes through the filter, or else the filter blocks it. A system which has passed through the filter B_χ has no memory of the state it was in before it hit the filter--χ could have been obtained from any state at all (except for those strictly orthogonal to χ).

If ψ and χ have unit length, then the "probability of transition" from ψ to χ is denoted $|\langle\chi|\psi\rangle|^2$, and the scalar product $\langle\chi|\psi\rangle$ is called the transition amplitude. Since the phases of χ and ψ are not determined, neither is the argument of the complex number $\langle\chi|\psi\rangle$; only the difference between the arguments of, say, $\langle\chi_1|\psi\rangle$ and $\langle\chi_2|\psi\rangle$, has a unique meaning. The square of the modulus of the sum of two complex numbers depends not only on the numbers themselves, but also on the angle between them, i.e., the difference of their arguments. This results in "interference of amplitudes".

The interaction of a system ψ with a filter B_χ is a special case of what in quantum mechanics is called an observation or a measurement. A more general scheme is obtained if we suppose that the system ψ encounters a set of filters $B_{\chi_1}, \ldots, B_{\chi_N}$, where $\chi_1, \ldots, \chi_N$ are a complete set of orthogonal basis vectors; these filters should be imagined "in parallel" (in the sense of electric circuitry), so that the system can pass through any one of them and end up in a state χ_j. With such a set of filters one associates a representation of some physical quantity B, which takes the values $b_1, \ldots, b_N$ in the states $\chi_1, \ldots, \chi_N$, respectively; we say that the act of measurement or observation gives the value b_j for B on the state ψ if ψ has passed through the filter B_{χ_j}. Mathematically, the system of filters (B_{χ_j}) or the quantity B is represented by a linear operator $\mathscr{H} \to \mathscr{H}$ which takes the vector χ_j to the vector $b_j \chi_j$ for all j. All such linear operators, which realize a dilatation of $\mathscr{H}$ with respect to the N mutually orthogonal directions (with real coefficients), are called observables.

As an illustration we give an idealized description of the Stern-Gerlach experiment on quantum measurement of the angular momentum (spin) of silver ions. The Hilbert space $\mathscr{H}$ corresponding to the spin degrees of freedom of this system is two-dimensional. Silver is vaporized in an electric stove; the ions are collimated (lined up) by a small aperture in a screen, and the resulting beam is sent between the poles of a magnet which creates an inhomogeneous magnetic field. The ions in the initial beam are found in all possible spin states, but upon passing through the magnetic field they have a tendency to "collapse" into one of two stationary states χ_+ and χ_-; these are traditionally called the states with spin 1/2 and spin -1/2. When the beam leaves the region of the field, these states turn out to be spacially separated

because of the inhomogeneity of the field.   The beam is
divided in half; thus, the magnetic field acts as a set of
filters.

We see that a quantum "observation", in its very essence,
has nothing in common with a classical observation, because:
(a) the act of "observing" almost inevitably knocks the
system off its phase trajectory;   (b) at best, the act of
"observing" allows one to register the new position of the
system on the phase curve, but not its position up to the
moment of observation, the memory of which is lost;   (c) the
new position of the system is only statistically determined
by the old one; finally,   (d) among the quantum "observables"
there are physical quantities (which in reality play a fundamental role) that do not correspond to any classical observables.

A comparison between the values of the quantum and
classical observables can only be very indirect.   For example,
to a quantum observable B one can associate its mean value
$\hat{B}_\psi$ on the state ψ (in the sense of statistical averaging).
It turns out to be equal to $\langle\psi|B|\psi\rangle$ (if $|\psi| = 1$).   (The
reader can take this symbol simply as a new notation.)   The
mean value $\Delta\hat{B}_\psi = \sqrt{[(B - \hat{B}_\psi^2)]}_\psi$ then measures the variance of
the values of B relative to the mean value $\hat{B}_\psi$ on the
state ψ.

Let B and C be two observable quantities, and let
$[B,C] = \frac{1}{i} (BC - CB)$.   One can show that the "Pythagorean
theorem" in $\mathcal{H}$ gives the inequality

$$\Delta \hat{B}_\psi \cdot \Delta \hat{C}_\psi \geqslant \frac{1}{2} \ | \ [B,C]\hat{}_\psi \ |$$

which is the mathematical expression of the Heisenberg uncertainty principle. It is most often applied to pairs of observables B, C for which [B,C] reduces to multiplication by ℏ. Then the inequality takes a more familiar form: $\Delta \hat{B} \cdot \Delta \hat{C} \geqslant \hbar/2$, independently of ψ.

We note that if $\mathscr{H}$ is finite dimensional, then there are no such pairs of observables. The uncertainty principle is usually applied to quantum analogs of pairs of classical observables (a coordinate and the projection of momentum on the corresponding axis, energy and time).

Combination of Quantum Systems

When we spoke about classical observations, we noted that attempting a detailed description entails the inclusion of the observable system S into a larger system (S,T). Therefore, we should expect that the unusual properties of quantum observations can be understood better once we examine the principles of a quantum description of a combined system.

The postulate of quantum mechanics that relates to this asserts that the state space $\mathscr{H}_{S,T}$ of the combined system is a certain subspace of the tensor product $\mathscr{H}_S \otimes \mathscr{H}_T$ (if S and T are infinite dimensional, then this product must be completed; we omit these details). What subspace of $\mathscr{H}_S \otimes \mathscr{H}_T$ must be taken is resolved on the basis of later postulates. For the time being, let us consider the case $\mathscr{H}_{(S,T)} = \mathscr{H}_S \otimes \mathscr{H}_T$.

Already the very formulation of the mathematical model exhibits the possibility of completely nonclassical connections between the "parts" S and T of the combined system. In fact, it turns out that for the overwhelming majority of the states of (S,T) it is impossible to indicate the state in which S and T are found "separately", so that the notion of "parts" turns out to have very limited meaning. Actually, $\mathscr{H}_S \otimes \mathscr{H}_T$ does contain decomposable states $\psi_S \otimes \psi_T$, where $\psi_S \in \mathscr{H}_S$, $\psi_T \in \mathscr{H}_T$. When the system (S,T) is in one of these decomposable states, we have reason to say that it consists of S in the state ψ_S and T in the state ψ_T. But even for the state $\psi_S \otimes \psi_T + \psi_S' \otimes \psi_T'$ such an assertion is untenable. Meanwhile, the superposition principle allows us to construct a large number of such states $\sum_i \psi_S^{(i)} \otimes \psi_T^{(i)}$. The set of decomposable state vectors has dimension $m + n$, and the set of all state vectors has dimension mn, where m and n are the dimensions of $\mathscr{H}_S$ and $\mathscr{H}_T$, respectively; thus, almost all the states of (S,T) are indecomposable. In the overwhelming majority of states of (S,T) the subsystems S and T exist only "virtually".

This type of nonclassical connection between the parts of a combined system frequently cannot be explained on the basis of classical notions that the connection between parts of a system is realized by an exchange of energy between them. In fact, in the fundamental case when two identical systems S and T are combined, the phase space of the combined system does not, in general, contain even one decomposable state. Let S be a fermionic elementary particle, such as

an electron, and let T be another particle of the same
type. Then $\mathscr{H}_{S,T}$ is the proper subspace of $\mathscr{H}_S \otimes \mathscr{H}_T =$
$\mathscr{H}_T \otimes \mathscr{H}_S$ consisting of the vectors that change sign when S
and T are interchanged--these are linear combinations of
the vectors $\psi_1 \otimes \psi_2 - \psi_2 \otimes \psi_1$. It is then easy to see that
each state of the system of two electrons is indecomposable.
In particular, there are no vectors $\psi \otimes \psi$ in the phase space;
in a popular presentation one says that two electrons cannot
be found in the same state. This is the reason for the
existence of stable atoms and, in the last analysis, for the
existence of the material world surrounding man. But in a
popular presentation it is almost impossible to explain the
"quasidecomposable" states $\psi_1 \otimes \psi_2 - \psi_2 \otimes \psi_1$. We can say that
one of the electrons is in the state ψ_1, while the other is
in the state ψ_2, but it's impossible to say which of them is
in which state.

Our natural language is in an even more hopeless position when it must explain the difference between two identical fermions and two identical bosons, where the phase
space $\mathscr{H}_{(S,S)}$ consists of vectors in $\mathscr{H}_S \otimes \mathscr{H}_S$ that are
symmetric with respect to permutation. That is, one tries to
explain that it is possible for two systems to be "different,
but indistinguishable" in two different ways (even one way
caused a lot of trouble for ancient philosophy).

This is a suitable place for a digression on "natural
language". In reality, our ideas of "classical" and "nonclassical" physics are very closely connected to our idea of
what can and what cannot be expressed adequately by simple

words. Here the state of affairs is not at all trivial. Not only a popularizer but also a working physicist often first tries to explain a new phenomenon, law or principle in a "rough and ready" way. But one must recognize the place of such an explanation. It strives: (a) to name and to summon from memory the corresponding fragment of the exact theory with mathematical formulas, structures, etc., in much the same way that the code for a command in ALGOL turns on the process of fulfilling the command, which also constitutes its meaning; (b) to switch on a process of free association, i.e., to aid in discovering that something is similar to something else; (c) to create in the brain a structure of intuitive notions on a subject, whose purpose is not to replace precise knowledge of it, but rather to form value principles and facilitate rapid estimates--to help us see what to seek later, in what direction to explore, what is probable and what is improbable. (In particular, this is the value of popularization for specialists in another discipline.)

We must again emphasize the following point: the semantics of a verbal description of some fragment of physics does not, in general, correspond to a complex of natural phenomena, but rather to a fragment of the theory, whose semantics, in turn, is made explicit by other fragments of the theory, operational prescriptions, etc. Nevertheless, the compulsion to interpret verbal expressions directly can turn out to be extremely fruitful. It was in this way that quarks were discovered: when it turned out that the space of certain internal degrees of freedom of a hadron splits into

the tensor product of three subspaces, it became tempting to
consider these three subspaces as the internal degrees of
freedom of three new particles which make up the hadron.
These particles are the quarks u, d, s (they "are discovered but are not detected in the free state").

Returning to the problem of quantum observations, we are
led to the conclusion that their mathematical model is nonclassical because, in the first place, it is a rough version
of a much more complicated model devised to explain the
interaction of a system with another system--a "device".
When we first discussed the meaning of the mathematical
apparatus of quantum mechanics, we particularly emphasized
that the device is macroscopic and there is no hope for a
complete quantum theory for its process of interaction with
the system. We are thus forced to replace the device by the
linear operator of the corresponding observable. Hence, the
logic of mathematical description leads us to the following
considerations, which as a totality are almost contradictory.
To the extent that the abstraction of an isolated quantum
system is sound, we need only one "observable"--the energy
operator--to describe it. However, we should not connect
this observable with the idea of measuring energy, since the
act of "measurement" requires an extension of the system.
The localization of a system in its phase space can also be
produced by "measurement" of other observables, but they are
coarse models of a combination that is inaccessible to complete description (system + device + energy operator of the
system/device). After an interaction with the device, the

system may lose its individuality, and the idea that it begins a new life at a point of a new phase curve in its phase space may lose all meaning. Finally, since even before its interaction with the device the system was a part of something larger, most probably at no time does it have the individuality which is needed for an adequate model. Thus, it seems there is no closed system smaller than the whole World.

After all of this it is a miracle that our models describe anything at all successfully. In fact, they describe many things well: we observe what they have predicted, and we understand what we observe. However, this last act of observation and understanding always eludes physical description.

At the end of the 1960's, optical shutters were designed for cameras which permit exposures lasting only ten picoseconds. In this fantastically short time a light ray in water travels a distance of 2.2 millimeters, and one can obtain a photograph of a short laser impulse in a bottle, while it moves inside of it. In the Bell Telephone Laboratories, where such photographs were first made, a drop of milk was added to the water, in order to increase the scattering of the light and make the trace of the impulse brighter. This drop of milk is a symbol of human participation in the world, where it is impossible to be only an observer.

4. SPACE-TIME AS A PHYSICAL SYSTEM

"I do not define time, space, place, and motion, as being
well known to all."   (I. Newton)

"...time and space are modes by which we think and not
conditions in which we live."   (A. Einstein)

We feel ourselves localized in space and lasting in
time, and, in the last analysis, almost all schemes of modern
physics describe events that are happening in the arena of
space-time. However, since the time of the creation of
general relativity theory and quantum mechanics, there has
been an increasing tendency to consider space-time as a
physical system in its own right. The principles for describing space-time still remain classical. The difficulties
of quantum field theory apparently point to a contradiction
between these principles and universal quantum laws.

The principal mathematical image of space-time is a
four-dimensional differentiable space-time manifold, which
for brevity we call the World. A point of the World is an
idealization of a very "short" and "small" event, such as a
flash, or the radiation or absorption of a photon by an atom.
Moreover, a point of the World is only a potential event;
such a point is "prepared to take" an event, but it "exists"
with or without it. The model for an event that is

71

concentrated in a small region of space but continues for some time, such as the life of an observer, a star, or even a galaxy, is a curve in space-time, called the world-line or history of an event. It is extremely important to learn to imagine the World of Becoming as a World of Having Become, i.e., to imagine the whole history of the Universe or a large part of it as a complete four-dimensional shape, something like the "tao" of ancient Chinese philosophy.

The introduction of temporal dynamics is the next step. It is realized as follows. In the World a space-time distance is defined between two nearby points $x^\alpha = (x^0, x^1, x^2, x^3)$ and $x^\alpha + dx^\alpha = (x^0 + dx^0, \ldots, x^3 + dx^3)$. Its square $ds^2 = \Sigma \, g_{\alpha\beta} dx^\alpha dx^\beta$ is a quadratic form in the differences of the coordinates of the nearby points. Here x^α is an arbitrary local coordinate system. A classical observer with his small laboratory of rulers and clocks can establish a local system of coordinates in which $x_0 = ct$; x_1, x_2, x_3 are Cartesian coordinates in the physical space of the observer, t is recorded on his clock. The metric in a neighborhood of the observer is close to the Minkowski metric $dx_0^2 - (dx_1^2 + dx_2^2 + dx_3^2)$. If we imagine the World to be populated by observers whose coordinates act in domains which cover the World, then the coordinates of events must be recalculated from one observer to another. But the space-time interval between two nearby points will be the same even when calculated by different observers. The speed of light c is used to convert temporal units to spatial ones; it is in such units that time is measured. The world-line of an observer is his own stream of time: the observer's atomic clock counts off the values of the integral $\int \sqrt{ds^2}$ along this world-line, i.e., its length. There is no physically meaningful "common time" of the Universe (although it can sometimes be introduced into special models of the World). It is not at all surprising that two curves in the World with the same beginning and end may have different lengths; this is even true on the Euclidean plane. Therefore, it is no wonder that two observers who synchronize their clocks and go separate ways discover that their clocks disagree when they meet again. A less familiar phenomenon is that, if two points of space-time can be joined by a world-line of an observer, then among such lines there is one which is the longest, but none which is the shortest (in the Euclidean plane just the opposite is true). This is a special property of the Minkowski metric, connected with the fact that it is not positive definite: the square of the interval between distinct points can be positive, negative or zero.

Observers with the longest world-lines, i.e., the fastest flow of their own time, are called inertial. From the point of view of general relativity theory, they fall freely in a gravitational field. Their world-lines are called timelike geodesics.

The second important class of curves in the World is the class of trajectories of particles flying with the speed of light, such as neutrinos. Along such lines the space-time interval vanishes identically--"time stops" (if the particle is massless, as was believed until recently). The geometry

of such lightlike geodesics determines what a cosmological
observer can observe and, more generally, what events in the
World can influence other events. In particular, it is in
terms of this geometry that one can give a precise formulation of the postulate that no signal can be propagated faster
than light.

The Minkowski World

The simplest and most important concrete example of a
World is the flat Minkowski World. It is a good local imi-
tation of most other Worlds at most points. In this World
there is an inertial observer whose coordinate system (x^{α})
covers the whole World and has metric identically given by
$(dx^0)^2 - ((dx^1)^2 + (dx^2)^2 + (dx^3)^2)$. We fix an origin of
reference--a point on the world-line of this observer. The
Minkowski World $\mathcal{M}$ becomes a vector space with our chosen
coordinate system, and this vector space structure does not
in fact depend on the inertial observer if we allow translation of the origin of reference. Therefore, the concepts
of a line, plane, and three-dimensional subspace in $\mathcal{M}$ have
absolute meanings. The linear transformations of $\mathcal{M}$ that
preserve the Minkowski metric form the Poincaré group, and
the transformations in the Poincaré group that fix some
origin of coordinates form the Lorentz group. These are basic
symmetry groups in all of physics; more precisely, they are
the basic symmetry groups of physical laws: none of the
points of the Minkowski World or the coordinate systems of
different inertial observers is preferred over the others,

and all coordinate formulations of the same law must be equivalent.

The set of points at distance zero from the origin of reference P forms the light cone C_P with equation $(x^0)^2 - (x^1)^2 - (x^2)^2 - (x^3)^2 = 0$. The timelike geodesics that pass through P are lines lying inside C_P, and the lightlike geodesics are lines on C_P itself, i.e., the generatrices of the cone. The cone consists of two halves, a "past cone" and a "future cone". The time along a timelike geodesic flows from the past half to the future half. The distinction between the two halves of C_P depends continuously on P; this expresses the *existence of a single direction of time* throughout the Minkowski World, *despite the absence of a single time*.

The point P and a timelike unit tangent vector at P form a model of the "instantaneous observer" in the Minkowski World. The vector indicates the direction of the observer's individual time. The subspace in $\mathcal{M}$ orthogonal to this vector is a model of the three-dimensional physical space of the instantaneous observer. Its metric (with sign reversed) is obtained by restricting the Minkowski metric.

The physical spaces of two different instantaneous observers are different, even if the observers are located at the same point of $\mathcal{M}$. They intersect the world strip of, say, a ruler at different angles. In some approximation such an intersection is the instantaneous observable form of this ruler. Therefore, it can have a different length for different observers--the space and time coordinates can compensate for one another.

How should one then imagine the observation of a distant object, say a star, in the Minkowski World? Suppose the observer moves on a world-line P, and the star moves on its world-line S. We imagine the past half $C_{P_0}^+$ of the light cone at P_0 as it moves along with $P_0 \in P$. It "sweeps out" a certain part of $\mathcal{M}$, a domain of the World which an observer could observe. At a given point P_0 the observer sees the star at the point of intersection of S with $C_{P_0}^+$ by means of the light ray joining $S \cap C_{P_0}^+$ and P_0. But we must still figure out how to determine the visible position of the star in the sky of the observer. The problem is that his sky "lies in his physical space E_{P_0}", and not in the Minkowski World $\mathcal{M}$, and the position of the star is modeled by a ray in E_{P_0}. In order to find this ray, we must project a ray in $\mathcal{M}$--the half-line from P_0 passing through $S \cap C_{P_0}^+$--onto the physical space E_{P_0}. This projection is orthogonal (of course, relative to the Minkowski metric).

Thus, it is convenient to distinguish the "absolute sky" at the point P_0--the base of the past half of the light cone--from the sky of an instantaneous observer at this point, i.e., the projection of the absolute sky onto the physical space of this observer. In classical cosmography, the sky can be imagined as a crystalline sphere of indeterminate radius; angular distances are defined between points of the sky, and the geometry of the sky is the same as the geometry of a rigid sphere. But the angular distances

between stars will be different for another observer; flying
with a very large velocity in the direction of the con-
stellation Orion, we see that it becomes squeezed together,
while the opposite celestial hemisphere is stretched (astro-
nomers call this aberration). Thus, the mathematical structure of the "absolute sky" is not the same as the structure
of a Euclidean sphere: angular distances in it do not have a
meaning independent of the observer. A detailed study shows
that the natural structure for the absolute sky is the
complex Riemann sphere, i.e., the complex number plane extended by a point at infinity which is treated indistinguish-
ably from the finite points. More precisely, the Riemann
sphere is the set of one-dimensional vector subspaces in a
two-dimensional complex vector space; equivalently, it is the
complex projective line CP^1. In particular, the natural
coordinates of a star in the sky are complex numbers. We
choose three reference stars and assign the coordinates 0, 1,
∞ to them. Then there is a simple procedure that enables us
on the basis of observations to associate to any fourth star
a complex number z, which will be observer-independent.

If the coordinates 0, 1, ∞ are given not to the reference stars, but rather to three points of the sky joined by
three orthogonal axes of the spatial coordinate system of an
observer, then, of course, the complex coordinates z and
z' of the same star can be different for different observers
at a given point of the World. But they are always connected
by a fractional-linear transformation of the form z' =
az+b/cz+d, where a,b,c,d are complex numbers independent

of the star, for which $ad - bc = 1$. The matrix $\begin{pmatrix} a & b \\ c & d \end{pmatrix}$ with determinant 1 is an unusual representation of the Lorentz transformation connecting two inertial coordinate systems at a single point of the World. In modern physics, however, this representation of the Lorentz group is considerably more fundamental than the usual 4×4 transition matrices.

Curved World

More general models of the World are different from the Minkowski World in several respects. First, the metric cannot be reduced to the form $(dx^0)^2 - (dx^1)^2 - (dx^2)^2 - (dx^3)^2$, even locally in the coordinate system of an inertial observer. Second, there may be no global coordinate system at all. Third, the metric of the World and the history of matter and fields in the World are not independent--the curvature of the metric is determined by matter, and, in turn, the metric imposes strong restrictions on the possible histories of matter; these constraints are described by the Einstein equations.

Geometrically, the typical ways the World curves can be perceived by studying the possible behavior of nearby time-like geodesics (local aspect) and light cones (global aspect). We shall try to give a verbal description of the effects of a very strong curvature, leading to the concept of a black hole.

We imagine a world-line S of a point mass m. With it one associates the characteristic space-time interval length $2Gm/c^2$--the so-called Schwarzschild radius. The points of the World that lie at a timelike distance from S at most equal to the Schwarzschild radius form the Schwarzschild tube $\bar{S}$ around S. Far away from this tube, the World is almost flat (if we disregard the influence of other matter). But inside the tube the World is so curved that for any point $P_0 \in \bar{S}$ the future half of the light cone C_{P_0} lies wholly within the Schwarzschild tube. The gravitational field of the mass m does not release photons emitted inside $\bar{S}$. However, the past half $C_{P_0}^+$ does not necessarily lie in $\bar{S}$-- the Schwarzschild tube can absorb external radiation.

We now consider a more realistic model, when the mass m is concentrated in a finite domain whose radius may be greater than the Schwarzschild radius. For example, the Earth has Schwarzschild radius of about 1 cm, and the Sun has Schwarzschild radius of about 3 km. In this case the Schwarzschild tube has no particular physical meaning. But in the process of evolution a star of sufficiently large mass can collapse under the influence of its own gravitation, so that at some point K_0 of the world-line of its center the radius of the star becomes comparable to the Schwarzschild radius, and then it decreases further. The segment of the Schwarzschild tube after the point K_0 will be a domain of space-time that is not accessible to an external observer. The four-dimensional picture of what an external observer sees will be approximately the following. The past half of the light cone of the observer at any observation point intersects the world tube of a star, but this intersection always occurs before the point K_0. In other words, the local time along the observer's end of the light rays will tend to infinity, but the

local time along the world-line of the star emitting the
light rays will tend to a finite limit corresponding to K_0
(we recall that local time is the length of the corresponding
world-line). According to the observer's clock, the collapse
lasts infinitely long.

We repeat once again how the four-dimensional picture
relates to the visible image in the sky of the observer. In
a curved World one more step is added to the already complex
process of translation. This step is the construction of a
flat World that is tangent to the curved World at the point
where the instantaneous observer sits. To obtain the visible
image of a star S in his sky, the observer must: (a) construct in the curved four-dimensional World the incident light
geodesic joining the observer's position to the world-line of
the star; (b) construct the tangent half-line to this geodesic in the flat World which is tangent to the curved World
at the point where the observer sits; (c) draw the tangent
to the world-line of the observer in the same flat World;
(d) construct the instantaneous physical space of the observer in this flat World; (e) project the tangent to the
light ray from the star onto this space.

Thus the "shadows of ideas" move on the wall of the Cave.

Spinors, Twistors and the Complex World

If we adopt the idea that a point of space-time is
an idealization of the classical notion of a "smallest
possible event", then we are inevitably led to the necessity
of considering still other geometric models as our knowledge

of such events increases. For example, the act of absorbing
a photon is far from being completely characterized by indicating the point of the World at which it occurs--we must
indicate the energy and polarity of the photon. Nor does the
position of an electron on its world-line completely determine its state--the direction of its spin must be indicated.

Although both polarity and spin are quantum-mechanical
internal degrees of freedom, it is remarkable that their geometric description is included in a very natural way in the
geometry of the World. Namely, the value of polarity or spin
at a point of the World is a ray in two-dimensional complex
space, or a point on the Riemann sphere CP^1. It turns out
that this Riemann sphere can be naturally identified with the
absolute sky of this point, which, as we explained before, is
also a Riemann sphere. Therefore, for each World $\mathcal{M}$ we can
consider an extended World $\overline{\mathcal{M}}$, a point of which is a pair (a
point of $\mathcal{M}$, a point of the absolute sky over this point of
$\mathcal{M}$). There is a natural mapping $\overline{\mathcal{M}} \rightarrow \mathcal{M}$, making $\overline{\mathcal{M}}$ into a
bundle over $\mathcal{M}$ with fibre CP^1. Here if we replace each
CP^1 by the two-dimensional complex space of which CP^1 is
the space of rays, we are led to the famous spinor space of
Dirac.

The world-line of a spin 1/2 particle, say an electron,
is naturally represented as a line in $\overline{\mathcal{M}}$, rather than in $\mathcal{M}$.
Rays of light are also lifted naturally to $\overline{\mathcal{M}}$: at each
point a ray determines a point on the corresponding sky,
namely the point at which it is directed; the set of such
pairs in $\overline{\mathcal{M}}$ is the image of this ray in $\overline{\mathcal{M}}$.

R. Penrose suggested that one study the remarkable space H which is obtained if each ray in $\overline{\mathcal{M}}$ is contracted to a point. Mathematically, H is the quotient space of $\overline{\mathcal{M}}$ by the equivalence relation defined by the rays in $\overline{\mathcal{M}}$. In order to better understand the construction of H, we analyze the case when $\mathcal{M}$ is the flat world of Minkowski. Then no point in $\overline{\mathcal{M}}$ lies in the intersection of rays (in contrast to $\mathcal{M}$), and each ray either does not intersect a given sky, or else intersects the sky in just one point. Therefore, each sky CP^1 is simply embedded in H without self-intersection, and some skies do not intersect each other. The simplest space in which one can pack many complex lines is the complex projective plane CP^2. But it cannot be a candidate for the role of H, since any two lines in CP^2 intersect. In fact, H lies in CP^3, complex projective three-space, the space of "projective twistors". To be sure, the skies over the points of the Minkowski World are not all the lines in CP^3, but only part of them, lying in a five-dimensional hypersurface (the whole of CP^3 is six-dimensional). It is very useful to introduce additional "ideal" points of the flat world $\mathcal{M}$, whose skies correspond to the missing lines in CP^3. We obtain the compact complex space-time of Penrose, denoted $C\mathcal{M}$; the coordinates of points in it can be arbitrary complex numbers and, moreover, there is an entire complex light cone "lying at infinity".

Can such an abstract construction have any relation to physics? It seems that the answer must be yes. One reason is the comparatively recent discovery of an analogy between quantum field theory and statistical physics. If the purely imaginary coordinate ict is replaced by a real one in the basic formulas of quantum field theory, then, roughly speaking, they become the basic formulas of statistical physics (the role of ict is played by the inverse of the temperature).

Geometrically, such a replacement corresponds to a transition from the Minkowski World with its indefinite metric to a four-dimensional Euclidean World with the "sum of squares" metric. Conveniently, this World fits into $C\mathcal{M}$; one need only rotate the time axis in $\mathcal{M}$ by 90°. All the other points of $C\mathcal{M}$ are obtained by interpolating between the Euclidean and Minkowski worlds, and, apparently, descriptions of many important phenomena admit an analytic continuation from $\mathcal{M}$ to $C\mathcal{M}$ or to part of $C\mathcal{M}$. A standard example is the interpretation of the "tunnel effect" of quantum mechanics in terms of the classical evolution of a system in imaginary time.

The "Penrose paradise" $H = CP^3$ (it is natural to use the word "paradise" to refer to a space which contains all skies, but in which nothing remains of space-time) has quite recently turned out to be very useful in the study of Maxwell's equations and their generalizations--the Yang-Mills equations, which, we now believe, describe gluon fields connecting the quarks in a nucleon. There are profound physical reasons to think that a World filled only with radiation (or with particles traveling at velocities close to the speed of light, almost along light cones) can be described better in terms of

the geometry of H than in the real four-dimensional geometry
to which we have become accustomed. What binds us to space-
time is our rest mass, which prevents us from flying at the
speed of light, when time stops and space loses meaning. In
a world of light there are neither points nor moments of
time; beings woven from light would live "nowhere" and
"nowhen"; only poetry and mathematics are capable of speaking
meaningfully about such things. One point of CP^3 is the
whole life history of a free photon--the smallest "event"
that can happen to light.

Space-Time, Gravitation and Quantum Mechanics

The most important lesson that one learns by studying
the relations between our models of the classical World and
the quantum mechanical principles for describing matter, is
that we understand these interrelations very poorly--the
basic rules for the descriptions do not agree with one another.

Here is one example. If we try to mark a point of the
World on the world-line of a particle of mass m, we cannot
do this with error less than the Schwarzschild radius $2Gm/c^2$
of this mass. Therefore, to increase the accuracy we should
make use of particles whose mass is as small as possible. On
the other hand, as noted by Landau and Peierls, the uncertainty principle for (coordinate, momentum) and the fact that
all velocities are bounded by the speed of light imply that
the indeterminacy of position cannot be less than $\hbar/mc$; that
is, an increase in accuracy requires the use of particles
whose mass is as large as possible!

Both limits on the accuracy are equal for a mass which
is found from the equation $\hbar/mc = 2G/c^2$, i.e., precisely the
Planck mass. This common limit on the accuracy of position
measurement is the Planck length. Hence, the Planck unit of
length (recall that it is of the order of 10^{-33} cm) gives
the lower bound under which the notion of a space-time domain
is automatically inapplicable. A smaller domain is incapable
of supporting an elementary event. The events studied in
modern accelerators occur in considerably larger domains, but
nevertheless the limit of localization is clearly indicated
within the framework of contemporary theories.

Quantum principles interfere with the notion of a point
of the World as an elementary event in many other ways as
well. A free quantum particle with fixed four-dimensional
momentum k is nowhere localized--it is "uniformly spread
over the World", since its psi-function is the de Broglie
plane wave e^{ikx}. Two identical quantum particles must be
described by a psi-function depending on two points of the
World which is symmetric for bosons and antisymmetric for
fermions. But this literally means that nothing can "happen"
at a single point of the World; more precisely, a point is
inseparably connected with all the other points through the
psi-functions of the matter and fields.

In a future theory the image of the real four-dimensional World with the Minkowski metric may turn out to be
something like a quasiclassical approximation to an infinite
dimensional complex quantum World. For example, in geometric
optics, which is an approximation to wave optics, there is

the concept of a caustic: the set of points where the intensity of radiation in this approximation is infinite. It is tempting to think of the four-dimensional World as a kind of caustic manifold for an infinite dimensional quantum wave picture. Our difficulties with, for example, the infinite density of vacuum energy in quantum electrodynamics, would be solved nicely in this scheme.

The Lorentz group is a strange group from the real point of view, but if it is replaced by SL(2,C), a group of complex 2×2 matrices, then we obtain a very natural object--the group of symmetries of the simplest imaginable state space of a quantum system. Does this not mean that the spin degrees of freedom are more fundamental than the space-time degrees? The mysterious separation of the World into space and time is implicitly contained in the group SL(2,C), and therefore its existence is "explained" on the basis of rules that do not assume such a separation *a priori*. Moreover, as we have seen, a World without rest mass matter can be obtained from SL(2,C) (or its generalization SL(4,C)) without introducing space-time. The points of our four-dimensional World, or small regions in it, are distinguished by events which happen with rest mass matter. Perhaps, both mass and space-time result from a spontaneous violation of the symmetry of the basic laws.

It is difficult to invent such a theory. We are still trying to quantize the classical Universe as we did for the hydrogen atom, rather than trying to obtain its form as the limit of a quantum description. Perhaps the first successful quantum model of the World, say, near the Big Bang, will be quite simple mathematically, and only the habits of an inert mind hinder us from guessing it now. I would like to live to see the time when such a model is proposed and accepted.

5. ACTION AND SYMMETRY

"Physics is where Action is." (Author unknown)

A small ball rolling in a trough under the action of its
own weight is the simplest physical system. The rest state
is the simplest history it can have. The ball can rest only
at those points of the trough where the tangent is horizontal;
otherwise it rolls down an incline. We give the position of
the ball by a horizontal coordinate x, and we let V(x)
denote its potential energy (proportional to the height of
the trough) at the point x. The points x at which the
ball can rest are the solutions of the equation $dV/dx = 0$ or
$dV = 0$; the increment of the potential energy function as we
move a little away from such a point is very small compared
to the increment of the coordinate. At these points V is
stationary.

A second example is a soap film that is stretched
between two wire loops. The equilibrium form of a film at
rest is determined by requiring that for small deformations
its surface tension energy V varies by an amount which is
very small compared to some natural measure of the amount of
deformation. The surface tension energy V is proportional
to the area of the film; hence, the form of a film at rest is

88

a state in which the area is stationary. The function (or functional) V in this case is defined on a domain consisting not of numbers x, but of all possible surfaces spanning the given contour. Instead of the differential dV we usually write the "first variation" δV.

For example, suppose that circular wire loops of radii r and R are located in parallel planes at a distance ℓ, and the line joining their centers is perpendicular to these planes. This line is an axis of symmetry of the contour, and therefore one would expect the equilibrium form of the film to be a surface of revolution of a curve q(x) for which q(0) = r, $q(\ell) = R$. We assume this to be the case; then the energy V is proportional to the area $2\pi \int_0^\ell q(x) (1 +$ $+ (dq/dx)^2)^{1/2}$ dx;   the equation $\delta v = 0$ can be rewritten: $-q(d^2q/dx^2) + (dq/dx)^2 + 1 = 0$;   and then we can solve this equation with the indicated boundary conditions.

Such a transition to formalism is both useful and dangerous. We postulated that the symmetry of the boundary conditions leads to symmetry of the equilibrium form. Here is a completely analogous example where this is obviously not the case. We load an elastic vertical rod with a compressing force; for some critical value of the load the rod buckles. The direction of buckling in a plane perpendicular to the rod is in no way distinguished in the original axisymmetric picture, but it is distinguished when the buckling occurs. In such situations physicists speak of a spontaneous viola- tion of symmetry: a phenomenon subject to certain laws is less symmetric than those laws. Besides, in our previous problem we neglected the solution consisting of two flat films stretched over each loop separately.

This is a reminder of the need, when studying a functional V on an infinite dimensional manifold such as the space of surfaces, to try to analyze the geometric situation beforehand. Topology is concerned with such problems. Only in recent years have the geometric ideas furnished by topology been systematically applied in physics, for example, the notion of "topological charge" has appeared in quantum field theory. (Unfortunately, there is no space for us to discuss this here.) Mathematically, it is useful to take $\delta V = 0$ as the basic equation. The domain of definition of the function V is perhaps not completely defined, but it may be made more precise in the course of the solution of the problem. The graph of V is an infinite dimensional trough, in which our system moves, seeking rest.

Physically, this image is justified by an amazingly universal principle, which can be stated as follows: *the development in time of fundamental classical systems is their equilibrium in space-time.* More precisely, the kinematics of a system is determined by describing its set of virtual histories in the appropriate space-time domain. A functional S having the dimension of action is defined on these virtual histories. The dynamics of the system is described by the condition $\delta S = 0$. The passage from the dimension of energy (V) to the dimension of action (S) is connected, of course, with the addition of the time coordinate. Action is primary; energy is only its time derivative. In the resulting fundamental theory action remains, while energy becomes a quasiclassical quantity.

The virtual history or path ϕ of a system in a space-time domain U is defined classically as a cross-section of the degree of freedom bundle of the system over this domain. If the domain is a union of two disjoint pieces U_1 and U_2, and the path ϕ is a union of the paths ϕ_1 and ϕ_2, then the action of ϕ is the sum of the actions of ϕ_1 and ϕ_2. This postulate is satisfied by almost all the action functionals used in classical physics. Therefore, the complete action of the path ϕ in the domain U can be written as a sum of actions of ϕ on many small domains which cover U, and in the limit as an integral $S(\phi) = \int_U L(\phi(x,y,z,t))$ dx dy dz dt, where (ϕ) is the set of intrinsic coordinates of the system and their space and time derivatives. The function L (or its integral over the space coordinates) is called the Lagrangian of the system: it is the density of action. If we place two systems with Lagrangians L_1 and L_2 in a domain of space-time, then the Lagrangian of the combined system has the form $L_1 + L_2 + L_{12}$, where the third term is the "density of the interaction". The two systems do not interact if this term is zero.

Space-time itself ("vacuum") contributes to the Lagrangian a term proportional to the curvature of space-time. Therefore, space-time can be considered on the same basis as systems which include matter or an electromagnetic field.

The role of action in quantum physics was clarified in an extraordinary fashion by Richard Feynman, who based his discussion on earlier work of Dirac. In the absence of a fundamental quantum theory we are forced to formulate his ideas as a recipe for "quantization", i.e., transition from the classical description of some physical system to a quantum description. According to this recipe, one imagines that each classical history ϕ contributes to the quantum description of the history of the system, but it does so with its complex weight (phase factor) $e^{-iS(\phi)}$. (The action is, of course, measured in units of $\hbar$.)

We explain this in more detail. We fix the classical behavior of the system on the boundary of the domain U, say ϕ_1 and ϕ_2 at moments of time t_1 and t_2. From these conditions--"loops for soap films"--quantum theory determines not the classical history of the development from ϕ_1 to ϕ_2, but rather a complex number $G(\phi_1,t_1,\phi_2,t_2)$--the amplitude of the transition probability from the state (ϕ_1,t_1) to the state (ϕ_2,t_2)--the square of whose modulus in principle is observable or enters into other observable quantities. Feynman's prescription says that this amplitude is $\int e^{-iS(\phi)} D\phi$, where the integral is taken over the infinite dimensional set of classical paths joining (ϕ_1,t_1) and (ϕ_2,t_2); the use of $D\phi$ instead of $d\phi$ serves as a reminder of the infinite dimensionality: it is not a differential, but a "volume element"!

In the prehistory of integral calculus, an important place is occupied by the remarkable work of Kepler

"Stereometry of Wine Barrels" (see Vol. 9 of his Gesammelte Werke). Integrals that give the volume of solids of revolution used in commerce were calculated in this work at a time when the general definition of an integral had not yet appeared. The mathematical theory of Feynman's magnificent integrals, which physicists write in vast numbers, is not really far removed from the stereometry of wine barrels.

From the viewpoint of a mathematician, each such calculation of a Feynman integral simultaneously defines what is calculated, i.e., it constructs a text in a formal language whose grammar has not previously been described. In the process of such computations a physicist calmly multiplies and divides by infinity (more precisely, by something which, if it were defined, would probably turn out to be infinite); sums infinite series of infinities, assuming here that two or three terms of the series give a good approximation to the whole series; and generally lives in a realm of freedom unencumbered by all "moral standards".

It would hardly be possible to construct a consistent and applicable mathematical theory of Feynman integrals without progress in the understanding of physics. The very idea of "quantization" belongs not to physics, but to the history and psychology of science. There can only be a substantive meaning to "dequantization", i.e., the passage from a quantum description to a classical one, when the latter is meaningful; the converse, however, can never have substantive meaning.

The classical fields occurring in the Lagrangians of the weak and strong interactions, for example, are physical phantoms: we do not know their meaning outside of quantization, and it is unlikely that they would describe the virtual classical histories of anything whatsoever. (It is believed that the situation is better with the quantization of electromagnetism.)

Finite ·dimensional quantum models allow us to guess which features of the Feynman formulation are essential, and which are atavistic. As was explained in Chapter 3, the evolution operator for a closed localized quantum system at its local time t has the form $e^{-iS(t)}$, where $S(t)$ is now an operator with the dimension of action. If we imagine the different world-lines of the system with different local times, we see that quantum action is a connection in the space of internal degrees of freedom of the system, determining physically admissible histories as parallel translations. The recipes for quantization are a primitive manifestation of the fact that the space of internal degrees of freedom "at a single point" *in vacuo* is already infinite dimensional because of the virtual generation of particles. Further understanding is blocked until we relinquish the idea of space-time as the basis for all of physics.

> "Symmetry denotes that sort of concordance of several parts by which they integrate into a whole. Beauty is bound up with symmetry." (H. Weyl)

Other than this quote, I shall not repeat anything from what Hermann Weyl wrote in his remarkable book "Symmetry"

(Princeton, 1952). It should be read by anyone who wishes to travel the road from perceiving symmetry as a sensory datum (colors, ornaments, crystals) to understanding it as the deepest of physical and mathematical ideas.

The universal mathematical structure describing symmetry is a group G and its action on a set X, for example, the group S_n of all permutations of the numbers $(1,\ldots,n)$. The action is a mapping $G \times X \to X$ which sets up a correspondence between a pair (an element g of the group, a point x of the set) and an element gx of the set (the image of x under the action of g). All elements of the form gx for variable g constitute the orbit of x under the action of the group. The group G itself is never given as a physical object--we can imagine a rigid body as a sensory datum, but the set of all rotations of it is an idea located on a higher level of abstraction. The homonymy of the words "action" in the contexts "action of a group" and "action of a segment of a virtual history" in the basic European languages is a random consequence of the vague initial idea of "change as a result of doing"; but in the expression $U(t) = e^{-iS(t)}$ for a quantum evolution operator, this homonymy unexpectedly acquires a deep significance.

One of the great achievements of mathematical thought was the separation of the concept of an abstract group from the concept of a group action on a set; this has turned out to be of great importance for physics. We imagine a hydrogen atom as an electron, moving in a central Coulomb field around a fixed nucleus. The group $SO(3)$ of rotations around the

nucleus acts on the complex vector space of quantum states of
the electron. It turns out that the entire infinite dimensional space of possible un-ionized states splits into a sum
of finite dimensional subspaces, with SO(3) acting separately on each. These actions are irreducible linear representations of the group, and are the possible stationary
states of the hydrogen atom. The precise mathematical
description of these representations explains the spectrum,
quantum numbers, etc. Similarly, the Poincaré group--the
full group of symmetries of the Minkowski World--acts on the
space of quantum states of a solitary particle in an imagined
World containing nothing besides the particle. As Eugene
Wigner showed, the experimental classification of the elementary particles by their mass and spin is included in the
classification of (infinite dimensional) irreducible representations of the Poincaré group. This gave rise to the
joke that the twentieth century world consists not of the
four elements air, fire, earth and water, but of the irreducible representations of some group. In recent decades,
theoretical physics has been occupied with an intensive
search for the symmetry groups of fundamental interactions;
their laws (Lagrangians) appear as a secondary object in the
mathematical description.

Physical accounts that stress formalism rather than
set-theoretic interpretation are sometimes obscure about what
the group of symmetries acts upon. Here are two extreme
cases, which lead to fundamentally different physics:
(a) the symmetry group G acts on the phase space of a

system and takes its phase portrait into itself; (b) the
system's phase space X is itself represented as the set of
orbits of some other space Y under the action of the group
G.

Case (a) includes the description of the hydrogen atom
considered above. The discrete invariants of an irreducible
representation are the quantum numbers of the corresponding
state, but the state vector in the representation space is
itself not uniquely determined, as in the example of the
buckling of a rod, in which the symmetry is spontaneously
violated. In schemes for describing fundamental interactions,
one often postulates a scheme for the violation of symmetry
by means of a weaker interaction.

Violation of symmetry is also an ambivalent term. It is
possible to speak of the spontaneous violation of symmetry
when the phase portrait of the system is symmetric, but its
individual trajectories are not. Accounting for a new interaction generally violates the symmetry of the whole phase
portrait, since it changes the phase portrait. But if this
change is small, then its effect can be approximated by considering the new interaction as a small perturbation of the
original symmetric picture.

Case (b) includes the so-called gauge theories. A
classical matter field in such a theory (an object capable of
quantization) is not a cross-section of the bundle of internal
degrees of freedom, as we stated earlier, but rather an
entire orbit of such cross-sections under the action of a
gauge group of transformations. Over a single point of

space-time such a group is represented by certain rotations of the space of internal degrees of freedom, but these rotations can vary independently and continuously as the point varies.

The condition that the Lagrangian of the theory should be invariant relative to such transformations imposes very stringent requirements, which sharply limit the choice of Lagrangian. In supersymmetric theories, about which we spoke briefly in the first chapter, the gauge group can be a still more general object; in particular, it can mix up the boson and fermion fields. It is the gauge and supersymmetric theories which are now rekindling the basic hope for a Grand Unification, i.e., the construction of a unified theory of fundamental interactions.

In conclusion, I would like to say a few words about the theory of numbers, a mathematical discipline which is highly developed and possesses amazing beauty, but which has as yet found no deep applications to natural science. One of the principal objects of study in number theory is prime numbers (a prime is a positive integer that has no integer divisors besides itself and one). Euclid's *Elements* already contained a theorem to the effect that there are infinitely many prime numbers. (If we have a finite set $p_1, \ldots, p_n$ of them, we can construct another prime number as the smallest divisor of the number $p_1 \cdots p_n + 1$ besides one). This is a remarkable argument, for all its simplicity.

Here is a more modern example of a theorem on prime numbers. Let $\tau(n)$ (the nth Ramanujan number) denote the nth coefficient of the series that is obtained after a formal expansion of the infinite product $x \prod_{m=1}^{\infty} (1-x^m)^{24}$. *If* p *is a prime number, then* $|\tau(p)| < 2p^{11/2}$. It is completely impossible to prove this here; according to the author of the proof, P. Deligne, in order to present this proof, presupposing everything known by a beginning graduate student in mathematics, one would need about two thousand pages of printed text. This theorem probably holds the record in modern mathematics for the ratio of the length of its proof to the length of its statement. Of course, the proof has led to a better understanding of many interesting things. For example, in order to prove this theorem a vast new theory was created ("ℓ-adic cohomology") and two or three old ones were used (Lie groups, automorphic functions,...).

It is remarkable that the deepest ideas of number theory reveal a far-reaching resemblance to the ideas of modern theoretical physics. Like quantum mechanics, the theory of numbers furnishes completely non-obvious patterns of relationship between the continuous and the discrete (the technique of Dirichlet series and trigonometric sums, p-adic numbers, nonarchimedean analysis) and emphasizes the role of hidden symmetries (classfield theory, which describes the relationship between prime numbers and the Galois groups of algebraic number fields). One would like to hope that this resemblance is no accident, and that we are already hearing new words about the World in which we live, but we do not yet understand their meaning.